Evidence from the Earth

Forensic Geology and Criminal Investigation

SECOND EDITION

Evidence from the Earth

Forensic Geology and Criminal Investigation

SECOND EDITION

Raymond C. Murray

2011
Mountain Press Publishing Company
Missoula, Montana

Photographs and other graphics not otherwise credited are by the author.
Cover image by Robert Rath

Library of Congress Cataloging-in-Publication Data

Murray, Raymond C.
Evidence from the earth : forensic geology and criminal investigation / Raymond C. Murray. — 2nd ed.
p. cm.
Includes bibliographical references and index.
ISBN 978-0-87842-577-8 (pbk. : alk. paper)
1. Forensic geology. I. Title.
QE38.5.M87 2011
363.25'62—dc23
2011024720

PRINTED IN THE UNITED STATES

P.O. Box 2399 • Missoula, MT 59806 • 406-728-1900
800-234-5308 • info@mtnpress.com
www.mountain-press.com

To Maureen, Elaine, Robert, Donna, Martha, Dick, Ellen, John

Contents

Acknowledgments

Many people over the years have been kind, thoughtful, and helpful and have provided opportunities and ideas to me in the forensic field. I will mention a few, realizing that the list is terribly incomplete: Nehru Cherukupalli, Jack Crelling, Laura Dawson, Laurance Donnelly, James Donovan, Nelson Eby, Chris Fiedler, Alexandra Guedes, Bruce Hall, Yoshiteru Marumo, Carlos Martin Molina-Gallego, Skip Palenik, Alastair Ruffell, Richard Saferstein, John Tedrow, Pornsawat Wathanakul, and Jack Wehrenberg. I am especially indebted to FBI forensic geologist Maureen Bottrell for providing me with her extensive expertise and brilliant insight into the issues of the science. In writing this book many new and old friends have provided me with cases, illustrations, and encouragement for which I am most appreciative. They include David Abbott, Sarah Andrews, Elisa Bergslien, Richard Bisbing, Bill Booth, Debra Croft, Clark Davenport, Rosa Maria DiMaggio, Howell Edwards, Rob Fitzpatrick, Olga Gradusova, Tom Hopen, Donald Hyndman, Wayne Isphording, Janet Kirkwood, David Korejwo, Tim Ku, Ian Lange, James McQuillan, Frederick Nagle, Ekaterina Nesterina, Gianni Lombardi, Leonardo Nuccetelli, Jim Oberhofer, Richard Olsson, Ken Pye, Ron Rawalt, Bill Schneck, Andrew S. Smith, Marianne Stam, Ritsuko Sugita, Graham Thompson, Erwin Vermeij, Margaret Waters, Jodi Webb, Patricia Wiltshire, and Andrew Wolfe. Needless to say, I am indebted to all the good and talented folks at Mountain Press for making this book a reality.

Preface

One of the interesting things about life is that you never know when a small event will have a major impact on your future. This happened to me one morning in 1973. Special Agent Ronald Decker of the Bureau of Alcohol, Tobacco, and Firearms walked into my office in the geology department at Rutgers University with several bags of rocks and soil. These bags held evidence related to two New Jersey criminal cases. In one bag was rock that had been used to break a window. It appeared that offshoots of two political terrorist groups were breaking one another's windows with rocks and then tossing in lighted matches and bottles of Coleman fuel. They might have been trying to make a statement that the Mob did not have a monopoly on violence in the Garden State or more likely they just hated each other. The rocks were picked up locally and, thus, were of little use in identifying the arsonists. The other case was an explosion at a plant that manufactures smokeless powder. The question asked was whether the explosion was an accident or sabotage. As a geologist, I was able to provide some information that could guide the investigation.

Ron had introduced me to a world of geology that I did not know existed. In addition, he introduced me to Richard Saferstein, then chief chemist of the New Jersey State Police Laboratory. Since that time, Saferstein, who is the author of the world's leading book on forensic sciences, has been a wonderful and generous friend. Saferstein's *Criminalistics* is now in its tenth edition. Ron Decker and his wife now operate a successful private investigation agency specializing in arson and bombings. Richard Saferstein introduced me to the forensic geologists at the laboratory of the FBI in Washington. A collection of highly skilled scientists totally dedicated to their profession and the law have worked in that laboratory. Among those who have offered me their time, ideas, and friendship are

Richard Flach, Elmer Miller, Cris Fiedler, Bruce Hall, Ron Rawalt, Maureen Bottrell, Jodi Webb, and David Korejwo.

One of the world's great soil scientists, Rutgers professor John C. F. Tedrow, has worked on many cases involving soil evidence. Together we published *Forensic Geology: Earth Science and Criminal Investigation* in 1975. In 1991, we updated it and retitled it *Forensic Geology. Evidence from the Earth* includes material from those two books.

It amazes me to see what has happened since the initial appearance of *Forensic Geology*. The book has been introduced in innumerable trials, both criminal and civil. Methods forensic geologists used thirty years ago and that we criticized in our books are no longer used. The number of cases involving soil evidence has increased because evidence collectors have become aware of the value of this evidence. The quality of the work of those who examine soil evidence has improved dramatically. Professors have developed college courses in forensic geology, such as the very successful one at Southern Illinois University under the direction of award-winning coal geologist Jack Crelling and the one at University of Massachusetts taught by Nelson Eby. There is now an organization known as the Geological Society of London Forensic Geoscience Group. Laurance Donnelly, an internationally famous searcher of buried bodies and objects, organized and developed this group with the assistance of many practitioners. They organize and hold international conferences on forensic geology and provide a means for forensic geologists around the world to communicate. I was honored and humbled to receive their first award for contributions to the profession in 2010. Donnelly's efforts have now resulted in the creation of the International Association of Forensic Geologists sponsored by the International Union of Geological Sciences.

Forensic science has captured the public imagination. Television programs, magazine articles, and radio bombard us with high-profile cases and the latest happenings in the crime lab. We watch *Forensic Files* for great reenactments and *NCIS* for their wonderful forensic scientists who can do anything. *CSI* gives us fiction and has had the impact of creating the CSI effect. The effect is at least twofold: Juries expect physical evidence and become disturbed when there is none, and crime lab storage facilities become filled with collected evidence in order to not disappoint juries. Everyday conversations touch on the latest achievement in physical evidence or the most recent court decision on admissibility of evidence. The

O. J. Simpson case and the attempted bombing in Times Square in New York in 2010 and the actual bombing of the London rail system in 2005 have turned us all into forensic science junkies. Even writers of fiction have turned to forensic geology for themes and plots. For those who like great crime novels, Sarah Andrews produces some of the best. Her heroines, forensic geologists Emily Hansen and Valena Walker, quietly move from book to book helping to solve horrible crimes and keep the reader coming back for more. Susan Cummins Miller has forensic geologist Frankie MacFarlane solve crimes with skill and intuition in wonderful stories. There is now a book on forensic geology for young readers published by Capstone Press titled *Earth Evidence.*

I hope this book will introduce you to the world of evidence locked in earth materials, show you the value of that evidence, and walk you through some of the cases from around the world that have benefited from forensic geology. It is an exciting and rewarding activity.

1 Forensic Geology in the Headlines

THE MURDER OF JOHN BRUCE DODSON produced one of the most interesting cases in the entire history of forensic geology. The geologic evidence is unequivocal: it tied the suspect directly to the crime and eliminated the suspect's alibi. Most important, the investigator of the crime recognized the potential importance of the geologic evidence and arranged for its examination. The testimony of the forensic geologist was critical to the prosecution of the case.

The case began on October 15, 1995, when Janice Dodson, John's wife of three months, shot and killed her husband on a crisp autumn morning during a hunting trip high in the Uncompahgre Mountains of western Colorado. Some would say that the case really began three months before when Janice started accumulating life insurance on John and making other financial arrangements to her benefit in the event of John's death.

At first glance it appeared to be a hunting accident. However, the autopsy revealed two bullet wounds to the body and one bullet hole through John's orange vest. Colorado District Attorney Frank Daniels points out in his book on the case, *Dead Center,* that if there had been only one bullet, there never would have been an investigation and the death would have been ruled an accident.

The Dodsons were camped near other hunters, one of whom was a Texas law enforcement officer. He responded to Janice's screaming call that her husband had been shot. She was standing about 200 yards from the camp in a grassy field along a fence line. The officer determined that John was dead and started the process of getting help.

Prior to calling for help, Janice had returned to her camp and removed her hunting coveralls which were covered with mud from the knees down. She later told investigators that she had stepped into a mud bog along the fence near camp. Investigators found a .308 caliber shell case approximately 60 yards from John's body. In addition they found a .308 caliber bullet in the ground, on the other side of the fence, in a direct line from the location of the shell case to the body and on to the bullet.

J. C. Lee, a former husband of Janice, was camped three-quarters of mile from the Dodsons. Janice knew this from past hunting trips as his favorite camp location. He naturally came under suspicion. However, he was hunting far away from camp with his boss at the time of the shooting. Most important, Lee reported to investigators that while he was out hunting someone had stolen his .308 rifle and a box of .308 cartridges from his tent.

Winter comes early at 9,000 feet in the Uncompahgre and little more could be done at the scene. However, investigators Bill Booth, Dave Martinez, and Wayne Bryant returned during the summers of 1996, 1997, and 1998 to search for the rifle and other evidence. They combed the entire area, including ponds, with metal detectors with hopes of finding the rifle. It has never been found. During the final search of the pond near Janice's ex-husband's camp, Al Beiber of Necro-Search commented that the mud in and around that pond was bentonite, a clay that someone brought to the pond to stop the water from seeping out the bottom. That evening Booth and Martinez were camped near the crime scene. They were discussing the evidence in the case when investigator Bill Booth said, "the Mud." He was referring to the dried mud that was found on Janice Dodson's clothing. If Janice Dodson had obtained the rifle from J. C. Lee's camp, she would most likely have stepped in or fallen into the bentonite clay that drained across the road from the cattle pond.

Remembering Janice Dodson's statement that she had stepped into a mud bog near her camp when she was returning there on the morning of October 15, 1995, Booth and Martinez decided they needed to obtain dried mud samples from the bog near Dodson's camp, the area around a pond near Dodson's camp, and the man-made pond and run-off near J. C. Lee's camp.

The mud samples that had been collected from Janice's clothing had been held at the sheriff's office evidence room since 1995. Booth and Martinez packaged the dried mud from each location and sent the samples along with the dried mud recovered from Janice Dodson's overalls to the laboratory section of the Colorado Bureau of Investigation in Denver for examination. Forensic scientist/lab agent Jacqueline Battles examined the material and later testified that the dried mud found on Janice Dodson's clothing was consistent with the dried mud recovered from the pond near J. C. Lee's camp. It was not consistent with the dried mud from the bog or the pond near her camp. This was a breaking point in the case. Battles is a highly respected forensic scientist with considerable geologic training who, like many others in the profession, got her early training with Walter

Pond where Janice Dodson fell and got bentonite mud on her shoes and jeans
—COURTESY OF WILLIAM G. BOOTH

McCrone, a pioneering scientist in the field of forensics. Her discovery allowed Booth and Martinez to place Janice Dodson in her ex-husband's camp around the time his rifle had been stolen. There are no other bentonite-lined ponds in the area and no bentonite deposits.

Bill Booth and Dave Martinez went to Texas and served an arrest warrant on Janice. She was extradited to Colorado, tried, and convicted of the murder of John Bruce Dodson. She is serving a life sentence without the possibility of parole in the Colorado State Prison for Women. The jury understood the results that followed Bill Booth's insightful exclamation, "the Mud."

WHILE IN MOST CASES chances are slim that a forensic geologist will find evidence as conclusive as that found in the Dodson case, the application of earth science to law has provided important evidence in countless trials around the world. Forensic geology is based on the principle, first stated by French criminalist Edmond Locard (1877–1966), that any time two objects touch, there is a transfer. You may not be able to detect it, it may be worn or washed away, but the transfer has taken place. So whatever people touch and whatever touches them leaves a trace. If you can find that trace, you can say where a person has been and possibly even place that person at a crime scene. Forensic geologists are interested in rocks, minerals, fossils, soils, and glass or other man-made materials or objects that have become incorporated in soil. Forensic geologists take samples of earth materials that have been transferred between objects and analyze them to determine their origins or sources. They then present the results as evidence in either criminal or civil legal proceedings.

The forensic geologist uses instruments and methods common to the profession of geology. Binocular microscopes, petrographic microscopes, X-ray diffraction, scanning electron microscopes, and microchemical analysis are examples of the forensic geologist's tools. The object of a forensic geologic examination is to identify and characterize the individual particles in a sample of soil or other earth material. The presence of rare and unusual minerals, rocks, or other particles can greatly increase the value of the evidence.

Forensic geology studies vary in scope. A common type of investigation involves identifying a material that is key to a case. For example, pigments in a painted picture or material in a sculpture are examined when authenticity or value is at issue. Identification is also important in questions of mining and mineral or gem fraud to determine if the material is what its sellers claim it to be. And identification of fire-resistant safe insulation on a person or individual's property may provide probable cause for further investigation.

Beyond identification, forensic geologists also look at the source of particular material, a task that requires a broad knowledge of geology and excellent geologic and soil maps. For example, if the soil on a body does not match the soil from the location where the body is found, the examiner will suggest locations that may match the soil on the body. Such studies are often called "an aid to an investigation." Examiners also compare two samples, one associated with the suspect and the other collected from the crime scene, to see if they had a common source: does the soil on the suspect's shoe have similar characteristics to the soil type collected at the crime scene?

Another new developing area of forensic geology is its use in intelligence work. A person, for example, may claim to have never been to a particular location but is then found with rocks from that spot, thus linking that individual to a geographic location. Remember the outcrop you saw behind Osama bin Laden on TV after September 11? What was the location? Geologist John Shroder, who had done field work in Afghanistan, was able to identify the region where bin Laden had been sighted in 2001. In addition, soil or dust on parts of a bomb may provide clues to the location of the bomb assembly area, or if different from bomb to bomb, the number of assembly areas.

Geologic evidence, as with all classes of evidence, rarely provides a truly unique solution for which the geologic mind cannot imagine another possibility. But there are some possible exceptions where the geologic evidence is especially strong.

So what does the word *compare* mean to the forensic geologist? We'll see this word many times in the context of forensics: A soil sample from the shoe of the suspect compares with soil samples from the scene of the crime; the rock found at the scene of the crime compares with rocks from a quarry in County Galway, Ireland. In examining the meaning a scientist places on this word, we must remember that no two physical objects can ever, in a theoretical sense, be the same. It is also true that a sample of soil or any other earth material cannot be said, in an absolute sense, to have come from a single place. Even when two samples are identical, there is always the possibility that similar material exists somewhere else on the planet and that therefore the two samples do not have a common source. Examinations of soil for forensic purposes seek to establish whether there is a high degree of probability that a sample was or was not derived from a given place. In practice, in determining if soil samples compare, the forensic geologist searches for the unusual, uncommon particles in samples.

The value of physical evidence generally depends on how many different kinds of that particular material exist. The more kinds and the fewer individuals of each kind, the greater the discriminating power and, thus, the greater the value of the evidence. Nature has provided us with many variations of soils, rock, minerals, and fossils. Every farmer and anyone who has taken a geology course knows that the number of kinds of rocks, minerals, and soils is almost unlimited. Furthermore, rock and soil type can change rapidly over short distances. This natural diversity of earth materials gives value to such evidence—in many cases, significant value. In fact, when we consider the value of evidence based on the number of variations, earth materials have greater useful value than almost all other forms of physical evidence, excluding such pattern evidence as fingerprints, tool marks, and DNA.

In considering such a statement, though, we must recognize three obvious limitations. First, soil evidence must have been present and legally collected by the investigator. This means in part that the investigator must be familiar with the possibilities for using earth materials as evidence and must know how to collect and treat samples. Second, the methods of comparison must be valid and performed by an expert at such a level of detail and thoroughness as to inspire a high degree of confidence in the meaning and usefulness of the comparison. If an untrained investigator examined only the color of a wet bulk soil sample, the soil evidence would have far less value than if an expert had examined many attributes of the sample. The third limitation is that a person may pick up earth material over time, making a comparison with samples from a single location meaningless. For example, soil from the floorboard of a vehicle may represent the accumulation from years of dirty shoes or from just one set of dirty shoes at one time. While it is sometimes possible to recognize distinctive and unusual minerals and rocks from a crime scene and compare them confidently with minerals and rocks removed from the floorboard of a vehicle, a superficial comparison of the two samples is meaningless.

Finding similar properties in two samples does not in itself prove a common source. The significance of such evidence is determined by probability and statistics. The advantage of soils, rocks, minerals, and fossils, or any physical evidence, lies in the scientific objectivity of the analysis and testimony of the expert.

As with any trace materials, soils must be collected if they are to be studied and presented as evidence. Whether a sample is collected depends on the knowledge and experience of the evidence collector. It is my hope that, in the future, all collectors of evidence will be knowledgeable and trained about the proper collection of soils in those cases where such evidence is present.

In this chapter, we will look at some legal cases in which geologic materials and methods contributed to the outcome. Some of these are high-profile cases, meaning that they received considerable press attention or that the victim or suspect was well known to the general public. Others illustrate interesting ideas, methods, or applications. In subsequent chapters, we will examine the basis for using geologic materials as evidence, explore how that is done, and finally look briefly at the future of the science.

Nine-year-old Rebecca "Becky" O'Connell lived with her parents in Sioux Falls, South Dakota. Her parents last saw Becky on the evening of May 8, 1990, when she left their home to buy candy at a nearby convenience store. Later that night, Becky's mother and stepfather reported her missing to the police. The following morning, two men found her body in a wooded area along a river in Lincoln County, South Dakota. An autopsy suggested she had been raped and had sustained knife wounds on several parts of her body. It was determined that she died as a result of a cut to her jugular vein. A man named Donald Moeller had been seen near her that evening, and this was reported to the police. Investigators examined Moeller's truck and collected soil from the fenders. They also collected soil samples from the area where the body was found. Forensic geologist Jack Wehrenberg examined the samples. In two of them, one from the vehicle and one from the crime scene, he saw unusual grains of a blue mineral. He thought they must be either blue diamonds or the zinc spinel called gahnite. A simple test excluded diamonds and the spinel was identified.

Gahnite is a rare mineral, and a search showed it had never been reported in South Dakota. Where did it come from? During the Ice Age, perhaps glaciers carried a decomposing cobble, or fragment, of a rock containing this mineral from Canada south to the spot where Becky's body was found. Whatever its origins, the presence of gahnite in both samples tied the suspect to the spot where Becky's body was found and strengthened the prosecution's case against him.

Moeller was convicted in 1992 of the brutal rape and murder of young Becky O'Connell and sentenced to death. The South Dakota Supreme Court reversed his conviction on procedural grounds in 1996, and Moeller was retried, reconvicted, and resentenced to death in 1997 by a Pennington County, South Dakota, jury. The South Dakota Supreme Court affirmed the conviction. In the second trial the defense hired a geologist with excellent academic credentials. The record shows that the heavy mineral separates sample—the sample that would include the gahnite—no longer existed at the time of the second trial. It is not clear what sample was given to the defense's expert. However, he took one grain from that sample and had a friend do chemical analyses on that grain. No zinc was reported, only silica and oxygen. The defense expert testified that the gahnite was misidentified and the grain was common quartz. He also stated that glacial soil cannot be used for forensic soil studies, which is certainly not true. The jury reaffirmed the original verdict.

A case that illustrates many of the issues regarding comparing soil and related material occurred in Canada. The body of eight-year-old Gupta Rajesh was found alongside a road outside of Scarborough, Ontario. The back of his shirt had a smear of oily material, and the preliminary conclusion was that he was the victim of a hit-and-run accident, with the oily material coming from the undercarriage of a vehicle. But examination of the oily material by forensic geologist William Graves of the Centre of Forensic Sciences in Toronto told a different story.

Investigators had collected samples of oily material from the floor of an indoor concrete parking garage where a suspect, Sarabjit Minhas, parked her Honda automobile. Analysis of the samples showed that the sand and other particles within the oil from the victim's clothes and the parking garage were similar. Oil collected from the floors of ten other garages in the area was also analyzed and compared to that on the victim's shirt and the suspect's parking garage floor.

Particles in samples from the victim's clothes and the suspect's parking place provided considerable information. The sand from both samples was sieved, and subsamples were produced of the various size grades from the two samples. When compared after the oil had been removed, the color of each pair of subsamples was identical. In addition, the heavy minerals in both samples were similar, and three distinct kinds of glass were found in the two samples: amber glass, tempered glass, and lightbulb glass. Each of the different glasses was identical in refractive index value (the amount a ray of light bends when passing through the glass into another medium). Small particles of yellow paint with attached glass beads were found in both samples. This type of paint is often found on center stripes of highways and reflects light.

Graves concluded that there was a high probability that the body of Gupta Rajesh had been in contact with the concrete floor of the garage at the place where the suspect parked her car. The same oil and particles were also found in the suspect's Honda. Whether the oil and particles on the victim came from inside the vehicle or the floor of the garage, the presence and distinctiveness of the samples strongly associated those two areas with the victim.

Minhas was tried in the Superior Court of the Province of Ontario in November 1983 and convicted, with help from testimony by Graves.

This case illustrates an important concept in the presentation of soil evidence and perhaps all physical evidence, except DNA. We have become awed and impressed by the high probabilities that result from DNA evidence. Some people expect that other types of evidence should have similar statistical information. But in the Minhas case, we see a conclusion based on at least ten different materials and observations. Because we do not know the probability of a tempered glass fragment, a particular group of heavy minerals, or sand of a particular color being on a particular parking place in a concrete garage in Scarborough, Ontario—and in all likelihood we will never know—a frequency statistic cannot be generated. A useful database of sands, particles, glass, oils, and heavy minerals would be too difficult to generate. In addition, it may not apply to any one specific case because of the variability of mineral particles—the very distinctiveness that makes geologic materials such good evidence. Thus, we rely on the skilled and honest examiner to reach a conclusion expressed in words rather than in numbers to inform the jury or judge so that they can reach a verdict. In this way the expert is a teacher, instructing the judge, attorneys, and jury in the basic concepts and premises that allow them to do the work they do. The triers of fact must be schooled in the methods of production of the evidence (how light bulb glass is made, for example), the procedures used to analyze it, and what makes the evidence significant. That understanding will lead the courts to an appreciation of unquantifiable evidence and give the jury a basis for weighing its significance.

FBI FORENSIC GEOLOGIST Chris Fiedler has worked hundreds of interesting cases, including the Philadelphia "Main Line" murder case made famous in Joseph Wambaugh's book *Echoes in the Darkness*. Investigation of the Main Line case led to several convictions, one of which was later overturned, for the murder of schoolteacher Susan Reinert and her two children. In this case, the presence of slag on the bumper of the vehicle that held the teacher's body led to a search of hundreds of Pennsylvania slag piles in the hope of finding her children's bodies. Unfortunately they were never found. However, the case illustrates an important method and the dedication of a first-class forensic examiner.

One of Fiedler's most interesting and important cases involved the search for terrorists who were moving explosives between safe houses in the Northeast during the 1980s. Investigators noticed that after a suspect vehicle passed a certain point in southern New Jersey, a large rock appeared by the side of the road at an intersection. In the coastal plain of southern New Jersey, you may find large grains of sand, but large rocks are rare. The rock must have been carried there. The investigators surmised that a lead car transported the rock and dropped it by the side of the road as a sign to the second car, which carried the explosives, that the coast was clear.

Investigators took the signal rock to the laboratory, removed a thin slice with a diamond saw, mounted it on a glass slide, and polished it to transparency. A polarizing microscope revealed the rock section to be garnet schist that contained an unusual form of the mineral staurolite. This was only the beginning. Fiedler contacted geologists at the Smithsonian Institution in Washington. To find out where the rock came from, they searched published descriptions and discovered that outcroppings of similar garnet schist occur in a very specific area in the highlands of western Connecticut. Investigators started asking questions in that area. Information led to more information and, ultimately, to a safe house in Pennsylvania filled with explosives. A smart, dedicated scientist followed clues in the rock to a place where human answers were available.

One case that attracted the attention of the world was the murder of Louis Mountbatten. On August 27, 1979, 79-year-old Lord Louis Mountbatten—distinguished war hero, diplomat, and elder statesman of Britain's royal family—was summering as usual at his turreted stone castle, Classiebawn, near the village of Mullaghmore, on Donegal Bay on the northwest coast of the Republic of Ireland. Around 11:30 a.m., Mountbatten, five family members and friends, and the "boat boy" climbed onto his 27-foot fishing boat, the *Shadow V*, and sailed out of the harbor. They checked some lobster pots and were sailing up the coast close to shore when a bomb blew the boat "to smithereens," as one of the locals described it. Fishermen rushed to the scene, but the blast had nearly severed Mountbatten's legs and he died almost immediately. Mountbatten's grandson, his daughter's mother-in-law, and the boat boy also died.

A security detail had been watching the house during the month each year that Mountbatten occupied it. However, they detected nothing. The boat had not been guarded and was moored with other small craft at a public dock. Two hours before the blast, Francis McGirl and Thomas McMahon, known members of the Provisional Irish Republican Army, had been detained at a routine checkpoint 70 miles away on suspicion of driving a stolen car. After the explosion, they were arrested and charged with the murders of Mountbatten and the others. Investigators collected trace evidence—sand and paint flakes—from the IRA members' boots and sent it to the Forensic Science Laboratory at Garda Headquarters in Phoenix Park, Dublin. The laboratory provides forensic services to the Garda Síochána, the national police of Ireland. The lab was then under the direction of Jim Donovan, known worldwide for the case in which a suspect on trial planted a bomb in an attempt on Donovan's life. Donovan suffered serious and lasting injuries. The event was portrayed in John Boorman's award-winning 1998 movie *The General*. In the Mountbatten case Donovan studied the sand that investigators collected from the boots of the suspects and determined that it was beach sand consistent with sand near where the boat had been moored. The paint also compared with paint from the boat. This evidence contributed to the conviction of McMahon for murder. McGirl was acquitted for lack of forensic evidence. At the time McGirl was originally detained on suspicion of possession of a stolen vehicle he declared, "I put no bomb on the boat." This statement was made prior to the explosion but was ruled by the court not to be a confession because he did not identify a particular boat. McMahon was released in 1998 under the terms of the Good Friday agreement.

A classic aid to an investigation occurred in southern Australia in the 1960s. Neighbors heard a struggle taking place at the home of two women. When police arrived, the women and their car were missing. The car broke down over 100 miles away. When the car was found, the suspect had returned to the car with a mechanic. The trunk of the car contained a shovel with attached soil. There was also a lot of blood. Rob Fitzpatrick, Director of the Centre for Australian Forensic Soil Science and an outstanding forensic geologist who has made major contributions to developing the field in Australia, examined the soil and the shovel. The soil on the back of the shovel had been smeared indicating the shovel had been used to pat down moist soil. Based on his knowledge of the local soils, Fitzpatrick determined that the soil on the shovel came from one of the local residual soil gravel quarries. The minerals in the shovel soil narrowed

Trunk of vehicle showing shovel with soil and blood spots in Australian double murder —COURTESY OF ROB FITZPATRICK

the search down to a small group of quarries. Samples from these quarries were collected and one produced identical minerals to the soil on the shovel. Rain had obliterated any sign of digging or footprints in that quarry. Police staked out that quarry and within three days a fox arrived, started digging, and unearthed the bodies. This led to the arrest and conviction of the man who came back to get the disabled car.

Not all cases result in such scientifically and judicially clear-cut answers. Much depends on the training, experience, and integrity of the forensic scientist and on the discovery of new methods of forensic examination.

In 1983 Albert Wesley Brown was convicted of murdering Earl Taylor in Wagoner County, Oklahoma. He was serving a life sentence when the Oklahoma Indigent Defense System's DNA Forensic Testing Program reexamined his case in 2001. Similar organizations in most states review capital and life sentences using DNA examination technology that was not available during the original trial. The DNA Forensic Testing Program determined that Brown's conviction rested on three primary types of physical evidence: fiber examination, microscopic hair comparison, and soil comparison. Fiber samples used in the original trial could not be found.

In attacking the state's case, the DNA Program examined mitochondrial DNA (mtDNA) from several hairs from the crime scene. The hair-comparison expert at the original trial opined that one hair found on the gag that had been placed in Taylor's mouth was consistent with Albert Brown's hair. The mtDNA revealed that this hair was not Brown's hair but a hair from the victim's own head. The expert also testified at the original trial that hairs found in the car Brown was driving were consistent with the victim's head hairs. These hairs were submitted for mtDNA testing, and testing revealed that they were not the victim's hairs.

Needing to consult a forensic geologist, the DNA Program asked me to review the soil comparison testimony from the original trial. Investigators had collected soil from the tires of Brown's car and examined it in the state crime lab. However, the agent who testified about the soil in 1981 had little formal education in geology, mineralogy, or soil science. During his testimony, for example, he mentioned glaciation in Oklahoma, when geologists know that Ice Age continental glaciers never reached that far south in North America. Unfortunately the agent had also collected the soil samples in the case using long-discredited methods. In the most seriously erroneous testimony, he stated that soils change systematically and that, within half a mile, they are similar.

This investigating agent had collected a control soil sample one-half mile from the crime scene. He stated that the sample was similar to the one collected from the suspect's tire and concluded that this placed Brown at the scene of the crime. However, soils and rocks do not change systematically. There are few places on earth where they are truly similar for distances of 10 feet, much less half a mile.

As a result of the Oklahoma Indigent Defense System's work, Albert Wesley Brown walked out of the Oklahoma State Prison on a motion for new trial. The state reopened its investigation and offered Brown time served with two years deferred if he would plead guilty to first-degree murder. Fearing a retrial, Brown accepted the state's offer.

Sometimes the finest work by the most skilled examiner leads to nothing but frustration and disappointment. On March 16, 1978, Italian Prime Minister Aldo Moro left his home in the northern part of Rome and headed for parliament. He had bodyguards in his car and more in the car behind him. Less than a mile into the journey, a Fiat 128 collided with the first car. The ambush had begun. Approximately ninety bullets were fired, five bodyguards died, and Moro was taken alive by members of the Red Brigades terrorist group. After fifty-five days, during which the government refused to negotiate with the terrorists, the group left a final message in the center of Rome: a red Renault 4 containing the recently murdered body of Moro. Moro's clothes and a blanket that wrapped his body yielded small samples of sand and plant fragments. The trunk, fenders, and tires of the Renault produced additional soil and related trace-evidence samples.

Famed Italian forensic geologist Gianni Lombardi was asked to examine the evidence. Lombardi provided his observations and conclusions to the court but waited twenty years to speak publicly about his findings in an article in the *Journal of Forensic Sciences.* What was the source of the sand? Examination of sand samples from the pant cuffs, shoes, and blanket showed them all to be similar. The grains were rounded and well sorted, suggesting an environment where the sand moved rapidly. Fragments of shells commonly found on beaches were present. Some of the grains on the shoes were stuck to dried spots of crude oil of the sort that sometimes washes up on the beaches near Rome. Lombardi thought the sand on Moro's body came from an area on a beach above the normal high-tide line. Study of the rock types in the sand revealed grains from outcroppings of metamorphic rocks along the coast north of Rome. Geologists also identified limestone with fragments of microfossils from an area north of Rome where rivers drain to the sea. Most interesting were fragments of volcanic rocks, some of which contained unaltered glass. These fragments had not traveled far. The source was also just north of the city.

Investigators undertook a massive study of beaches north and south of Rome. They sampled the sand from every beach accessible by car. Sand from Moro's beach house south of Rome was sampled and excluded as a possible source. Using all the information he had developed, Lombardi narrowed the possibilities down to a 7-mile stretch of beach, just north of the Rome airport, to which only a few roads led. Plant fragments found with the body were similar to those found in this area. Lombardi had done what every forensic scientist hopes to do: he had provided the investigators a real aid to the investigation. Police searched the area, unfortunately with no success.

Years later, investigators found a terrorist hideout in a suburb southeast of Rome. In their confessions, Red Brigades members said they kept Moro there and shot him in the garage just before driving the body downtown. Most disturbingly, they claimed they had collected beach sand and plant fragments and placed them on the body to confuse the investigation. Some experts question the plausibility of these confessions. It is possible that we will never learn the full story. Nevertheless, the case remains one of the finest examples of using forensic geology to search for the source of soil evidence.

In 2001, Lombardi was asked to examine evidence in a politically sensitive case. Countess Agusta plunged to her death from a cliff top near her forty-room villa located in the town of Portofino on the Italian Riviera. At 7 p.m. on January 8, 2001, the countess headed into the garden in her bathrobe and slippers and disappeared. The Carabinieri were called at 2 a.m. following an unsuccessful search of the house and grounds. Her body was later found approximately 200 miles to the west on the French coast. Was it murder, suicide, or accident? Murder was immediately eliminated because everyone on the premises was accounted for. There was a small area with a ledge and a tree on the top of the cliff in her front yard. This area had fragments of broken glass. Her slippers, which were found on the cliff, had fragments of the same kind of glass. The rest of her front yard was a direct drop-off down the cliff. Lombardi reasoned that if she was contemplating suicide she would have simply jumped off the cliff. The fact that she had stepped on the glass by the tree suggested she was standing by the tree on slippery leaves and fell by accident.

Olga Gradusova and Ekaterina Nesterina are forensic geologists with the Russian Federal Centre of Forensic Science. Gradusova is a graduate in soil science from Moscow University and Nesterina received a

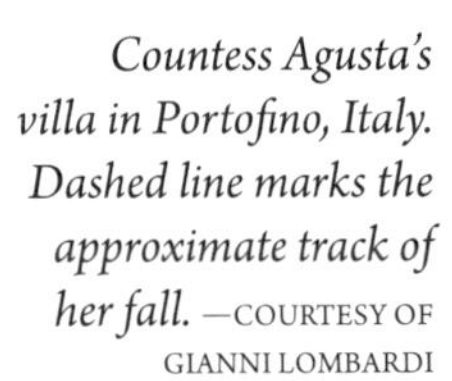
Countess Agusta's villa in Portofino, Italy. Dashed line marks the approximate track of her fall. —COURTESY OF GIANNI LOMBARDI

doctorate in chemical technology. They report a very interesting case that required a deep understanding of mineralogy. A teenager was raped in the wet basement of a house with a central heating system. The forensic scientists examined soil on the suspect's boots, soil on the victim's boots and clothes, and soil from the crime scene to determine whether both individuals involved had been present at the site. In this case, examinations of soil microparticles by light microscopy and other methods showed that the soil samples compared positively. The grayish brown soil included small fragments of glass fibers with iron hydroxide, glass fibers from a heating pipe insulator, aerosol ferromagnetic spheres from heating iron metal, small fragments of bones, and vivianite globules. Micro X-ray fluorescence spectrometry and electron microscope microprobe energy dispersive X-ray analysis were used to identify the vivianite, a mineral with the composition $Fe_3(PO_4)_2 \cdot 8H_2O$. The forensic scientists used natural samples of vivianite from a collection at the Museum of the Earth on the Moscow State University campus as standards. Most vivianite is found in natural rocks, but rapid growth of vivianite is possible, though rare, when phosphate, such as a bone, is brought in contact with iron fragments. The vivianite globules in the soil samples likely grew in the basement. The similarity of all samples, including their color and texture, and particularly the presence of vivianite, established that the suspect was likely at the scene of the crime.

Forensic geologists are frequently called in on crimes known as substitution cases. In these cases, people remove valuable cargo, such as guns, electronic instruments, or medication, from its container and substitute rocks or sand of the same approximate weight. At the container's final destination, the purchaser opens it to find only the rocks or sand. When a shipment has multiple insurers, where the substitution took place can determine who pays. Usually criminals do not carry sand or rocks far—they are heavy, after all—so investigators surmise that substitution took place at or near the source of the filler.

In one substitution case, a million dollars' worth of cigarettes were moved in containers by truck from Winston-Salem, North Carolina, to Norfolk, Virginia. There they were loaded on a ship that stopped in several South American ports, finally unloading in Montevideo, Uruguay. When the containers were opened, they contained bags of sand.

There were two insurers, one for the truck and one for the boat. If the sand was from South America, the boat insurer would pay. If it was from North America, the truck insurer would pay. How do you tell North American sand from South American sand? One way would be to travel the route, taking samples at each place the substitution could have occurred. This would be useful for focusing an investigation, but to simply solve the question of liability, pollen in the sand, from either South American or North American plants, would provide an answer. In this particular case, the study was never completed.

In a similar case involving missing computer equipment, the highly skilled forensic microscopist and president of Microtrace Inc., Skip Palenik, was called upon to examine evidence. The computer equipment had been shipped from Texas to Argentina via Miami. When opened in Buenos Aires, the crate contained only concrete blocks. Using gentle acids to dissolve the cement, Palenik removed sand from the concrete. It was fine with a very narrow distribution of grain sizes suggesting beach sand and a composition of heavy minerals suggesting a location in Florida. Palenik compared the sample with samples of Florida sand from his extensive collection of sands. They were similar. He told investigators that the people who substituted the concrete blocks probably did not move them far and to check the Miami airport. Investigators found identical blocks at a construction site at the airport. This led to the identification and conviction of those who made the substitution.

Palenik has worked on many high-profile cases including the Unabomber, the Oklahoma City bombing, the JonBenét Ramsey murder, and the assassination of Martin Luther King Jr. However, his most spectacular contribution came in the case of the Green River Killer, Gary Ridgway. Palenik's evidence caused the defense attorney, Tony Savage, to change his mind and plea bargain his client to a life sentence for forty-eight murders. The case began in 1982 with a rash of serial killings of prostitutes in the Pacific Northwest. Some of the early bodies were found in the Green River near Seattle, Washington. Ridgway was an early suspect, but the police did not have the evidence to hold him beyond questioning. DNA evidence on one victim led to his arrest in 2001. However, because the victim was a prostitute, the presence of Ridgway's DNA did not prove he murdered her. Ridgway worked painting Kenworth trucks and used a DuPont paint called Imron that is used in small amounts and mainly on Kenworth trucks. Naturally paint stuck to anyone working with the material, and Palenik found particles of this same paint on five of the victims. Ridgway had transferred it to his victims, and Palenik had found the connection between Ridgway and the serial killings.

In the Miami area, forensic geologist Fred Nagle stays busy investigating substitution cases. Containers from South America commonly arrive at both the port and the airport with substituted cargo. Nagle's favorite case took him to Brazil on the trail of a shipment that started out as 2 million dollars' worth of exotic perfume and arrived in Miami as several tons of sand. The shipment had traveled from Paraguay across Brazil to the port of Santos by truck, up the coast by ship to Rio and Salvador, and on to Miami. The investigation was limited to establishing liability—which insurance company should pay? The sand was fine-grained, rich with quartz, brownish, and in bags, some of them burlap and others made of a white modern synthetic fiber. Nagle took the bags and some of the sand to one of his graduate students, a native of Salvador, Brazil. The student immediately identified the bags as Brazilian coffee bags from the green and yellow thread sewn into their sides. The white bags were more modern coffee bags. The student knew that sand is generally sold loose in Brazil, not packed in bags. He also did not recognize the sand as coming from anywhere near Salvador, one of the freighter's stops on the way to Miami.

Nagle began his study by comparing the sand with Brazilian samples from the University of Miami's collection of sands of the world and some other collections. None of the samples was similar. Nagle decided to go to Brazil and check a few of the known freighter stops to look for the sand source. He was looking for a fairly mature, fine-grained, river-deposited, granitic sand that contained rutilated quartz. To prepare, he talked to other geologists and acquainted himself with the area's geology.

His first stop in Brazil was a small port south of Rio where the freighter had made an unscheduled stop. The freighter captain, who had been relieved of his duty, would not talk about why he made the stop. As captain, he had the right to stop wherever he wanted. However, his refusal to discuss the three-day stop made him a suspect. Nagle found that, at the port, access to the containers was easy. Anyone could walk around the truck containers without being stopped—further evidence that this might be the right place. But his search of the area for sand similar to that involved in the theft turned up nothing.

Nagle searched for two weeks. Finally, feeling defeated, he flew to Iguaçú Falls, on the border between Brazil and Paraguay, to see the falls and the new, giant hydroelectric dam there. He knew that basalt was the dominant rock in the region around the falls. An engineer friend of one of Nagle's colleagues showed him the dam. Nagle asked where they got the sand for the dam. The engineer said that most of it was ground-up basalt, but that some was dredged from the Paraná River. Nagle tested the river sand, and yes, it was a perfect match with the sand associated with the crime. There was a large pile of the sand at the construction site. As he crossed the top of the dam into Paraguay, Nagle also noticed sandbags. The engineer explained that they were used to keep people from falling and from getting too close to the construction. In the bags—coffee bags like those identified by Nagle's student—was sand similar in color and grain size to the sand from the construction site, sand from the flood terraces along the Paraná River, and the sand used in the crime. Laboratory examination of samples confirmed that they were identical to those from the substitution. Unusual grains helped make the identification more certain. All the sand samples contained fragments of blue turquoise. Blue turquoise fills the amygdules, or cavities, of the basalts in that area.

Nagle's investigation revealed that the substitution took place at the front end of the cargo's voyage. It helped many previously mysterious details fall into place, such as the source of the sandbags, why the container seals had looked undisturbed, and why the weight was correct—the weight was correct because the same cargo, the sand, had been there all along. Nagle's discovery also revealed the innocence of the prime suspect, the freighter captain. Nagle was later told that the captain had stopped to visit a girlfriend at the port, something he had not wanted to make public. That may or may not have been true, but after Nagle solved the case, he received a huge thank-you from the captain for clearing him and helping get his job back. It was later reported that several people were arrested in connection with this case. The stolen perfume had been diluted with water, put into bottles, and sold at the regular price, so that its original worth of $2 million became $6 million.

FBI forensic geologist Maureen Bottrell received two geology degrees from the University of Georgia. One of her most interesting cases took place in the southern part of that state. The town of Attapulgus lies just north of the Florida line. It is famous for an unusual rod-shaped clay mineral named attapulgite. Following a homicide near Attapulgus, police arrested a man in northern Florida. He worked in a cement factory and claimed that the white, powdery material on his clothes was cement and that he had not been to Georgia. Bottrell's examination showed that it was attapulgite identical to material from the crime scene in southern Georgia. Presented in court, this evidence contributed to the suspect's conviction.

Alastair Ruffell, one of the world's most insightful forensic geologists, teaches at Queen's University of Belfast and has been involved in cases throughout the world. An interesting case of his was a fatal gang-related shooting in the Greater Dublin Area in 2001. One gang led the leader of another to believe their joined forces could defeat a common enemy. The gang brought the rival leader to a secret weapons cache and discussed the division of arms and tactics. The following night the gang and the rival leader, who was no longer on guard, returned to continue the discussion. At the cache site, an abandoned quarry of Carboniferous limestone, a gunman appeared and shot the rival. The gang placed their victim in a nearby excavation, scattered some agricultural lime on top, and covered him with sandstone slabs, rubble, and soil. The lime was crushed limestone ($CaC0_3$), a slow-release acid soil improver. The victim was not reported missing for several days, but almost as soon as he was, a man walking his dog near the abandoned quarry saw the disturbed ground as well as some patches of clothing protruding through the earth.

Various suspects were arrested in connection with the murder, but there was little to connect them directly to the crime. At the gang leader's home, investigators found an expensive jacket that had light-colored smears on it, indicating the surface had been wiped and then dried. He had apparently destroyed his other clothes worn during the crime but could not part with the jacket. Detailed examination of the jacket indicated the light coating was calcium carbonate with no visible microbiological materials. An X-ray diffractometer established the mineralogy of the dust as predominantly calcite but including the mineral cookeite, a rare lithium-aluminium-hydroxy-silicate. Scanning electron microscopy also showed microfossils of Carboniferous age and a crushed or jagged appearance to the calcite crystals. The combination of a unique mineral, microfossils from a specific section of the Carboniferous rock, and the crushed nature of the calcite pinpointed the quarry as the source of the dust. The defendant admitted his presence in the quarry but pleaded to a lesser charge.

Another case where exceptionally strong geologic evidence helped an investigation began with a man taking a walk along the Missouri River outside Great Falls, Montana, following a path that Lewis and Clark had walked more than one hundred years before. He found credit cards and identification belonging to Susan Galloway and brought them to the Cascade County sheriff's department. Indeed, Galloway and her car had been reported missing. A further search of the river produced her wallet with other cards and one hundred dollars in cash. Then, on the side of a steep cliff above the river, investigators found the car itself, with Galloway's body in the trunk. She had been killed with a broken soft drink bottle and fragments of the bottle were still in her body. Blood covered the passenger seat and the outside of the car. Detective Sergeant Ken Anderson of the Cascade County sheriff's department believed that the crime had taken place somewhere else and that someone had pushed the car over the cliff. At a local lovers' lane, search dogs located disturbed earth where someone had attempted to dig a hole. The dogs had probably smelled blood. Lab tests later confirmed the presence of blood in the soil and identified it as Galloway's.

Galloway was engaged to Craig Smith, an officer at nearby Malmstrom Air Force Base. When Detective Anderson interviewed Smith, he saw that the young man had a fresh cut on his hand. Smith said it was from a knife that slipped while he was opening a can. Smith's story of the night his fiancée disappeared had many inconsistencies. According to the medic at the base, the cut on Smith's hand was not caused by a knife. A friend of Smith's at the base had seen him that night and driven him home. The friend's story differed from the one Smith told the police. In addition, a convenience store clerk remembered selling several soft drinks in green glass bottles to Smith and Galloway.

The only thing missing in the case against Smith was solid evidence placing him at various sites associated with the crime. Detective Anderson contacted forensic geologist Jack Wehrenberg and provided sneakers recovered from Smith's home plus soil samples from the various crime sites. Wehrenberg studied the samples. Smith had cleaned the sneakers, and only 50 milligrams of soil remained on the shoes. Still, Wehrenberg was able to ascertain that samples from the sneakers compared with samples collected at the site of the shallow digging where Susan's blood was found in the soil. Further investigation by Wehrenberg would considerably strengthen the identification.

A short distance from the digging site stood the former copper smelter and refinery of the Anaconda Copper Mining Company. For more than fifty years, the facility's large smokestack had emitted spherical glass particles into the air. With time, they became incorporated in the surrounding soil. These spherical glass particles are products of the smelting process; their exact composition and characteristics depend on the smelting process used and the kind of ore being smelted. The kind, color, and size of the glass particles also varies from place to place around the smelter, providing excellent markers in the surrounding soil. In addition to the spherical glass particles, soil samples from the shallow digging location contained fragments of brown, green, and clear container, or bottle, glass. The shallow digging site was the only place around the smelter where this combination of glass particles was found. Glass particles, both spherical and container, from the sneaker soil samples matched those from the digging site. The spherical glass particles, with their distinctive colors, made the comparison especially convincing.

The value of Wehrenberg's evidence was indisputable. Smith was convicted of the murder of Susan Galloway and sent to the Montana State Prison at Deer Lodge for one hundred years. This example of fine forensic work was featured on a forensic science television program called *The New Detectives*.

The Charles Moses case also turned on evidence stemming from the expert examination of glass. On February 16, 2000, terror gripped the small town of Paxton, Nebraska, and the surrounding areas. Schools were closed, the governor declared a state of emergency, and military helicopters filled the skies as more than one hundred law enforcement officers searched for Charles Lannis Moses, Jr. Three days before, a western Nebraska sheriff's deputy had attempted to arrest Moses on a Texas warrant for theft and evading arrest. In the ensuing confrontation, Moses wounded the deputy and a Nebraska state trooper. Two days later Moses killed a farmer in his field and stole his pickup truck. At the time he was shot, the farmer was talking on a cell phone to his father-in-law, who had called to warn him to watch for the fugitive. Moses abandoned his own Chevrolet pickup at the farm. Next he robbed an elderly woman at her home. The next day, witnesses spotted Moses in Wyoming driving the stolen truck. After a high-speed chase, Moses abandoned the truck and took off on foot. He entered a farmhouse and was captured by a citizen.

FBI forensic geologist Jodi Webb, who counts among her many accomplishments a mastery of glass examination, was asked to study a number of glass samples associated with the crime spree. Using the GRIM II (Glass Refractive Index Measurement) system for refractive index determination, visual examination with a microscope, and chemical analyses, Webb was able to make the following associations:

- Glass from Moses's truck, abandoned in the farmer's field, compared with glass from Moses's clothing.
- Glass from the driver's side of the farmer's stolen pickup compared with glass from the farmer's field, the farmer's body, and the location of the robbery of the elderly women.

These associations tied other evidence together and led to a guilty plea and life sentence for Moses.

Forensic geology can help uncover criminals even when they try to cover their tracks. Jennifer McCrady, the wife of Ohio State Trooper Jackie McCrady, disappeared from their home in Belpre, Ohio. A local citizen reported that a police car driven by someone resembling Jackie had been seen in a remote area outside of town the day Jennifer disappeared. At first, investigators paid little attention to the report, but when all other leads proved fruitless, they searched the remote area and found a mound of disturbed dirt and recovered Jennifer McCrady's body. She had been shot with a .357 Magnum revolver and buried with bedding from her home. Jackie became a suspect, and investigators recovered a .357 Magnum from his home. Someone had polished out the rifling in the barrel of the revolver, making it impossible to compare the markings on the bullet recovered from the body with bullets fired from that revolver.

An alert investigator had seen soil-covered shovels during a search of the McCrady home. Richard Bisbing, forensic scientist and vice president and director of research for the internationally recognized McCrone Associates of Westmont, Illinois, studied soil samples from the grave and from the shovels in Jackie McCrady's garage. Bisbing found that the samples from the grave and shovels were identical in color, grain-size distribution, and mineralogy, and that the samples from the shovel did not match samples collected around the McCrady home. Bisbing's testimony helped the jury reach the verdict that Jackie McCrady was guilty of murdering his wife. He was sentenced in 1997 to fifteen years to life in the Ohio state prison. Altering the barrel of his revolver slowed down the investigation, but, partly because of evidence gathered through forensic geology, it did not spare McCrady from being convicted.

EVERY ONCE IN A WHILE, a case develops in which multiple suspect-associated samples of different soils compare with samples from different sites associated with the crime. This naturally increases the evidential value. Andrew Wolfe, who examined the following case, is a forensic geologist with the Centre of Forensic Sciences in Toronto, Canada. The laboratory, which serves the province of Ontario, has contributed significantly over the years to the science of forensic geology. This particular case illustrates the importance of communication between the soil examiner at the laboratory and the investigating officer. In this case, such cooperation resulted in the collection of additional samples, greatly improving the value of the evidence.

The suspect, a young man, arranged to meet his stepmother in a local parking lot to discuss issues relating to an inheritance. Late in the evening, a resident in the neighborhood heard a commotion and witnessed a male run from the parking lot, cross the road, and ride away on a bicycle. The witness found the stepmother dead on the ground by her car. The suspect's running shoes and two soil samples collected from the middle of a large field adjacent to the parking lot were submitted for forensic comparison.

Both shoes had distinct soil deposits on the heel and the sole. Soil from the left heel had the same color, texture, and grain-size distribution as that of the two samples from the field. Mineral contents of the samples were also the same except the shoe sample contained a small amount of limestone, whereas the two field samples contained none. Three other soil deposits on the shoes also contained limestone, and differed in color and texture from the field samples, indicating contact with other limestone-containing soil sources. While investigators had some knowledge of the route the suspect took between his stepmother's car and his bicycle, there were no footprints to indicate the exact sites of soil transfer. Since the two field samples showed no significant variation in properties and differed from the soil on the left heel only in their lack of limestone, could the soil transfer have occurred at the edge of the field, where soil alteration is more common? Andrew Wolfe required more information about the scene in order to assess the significance of the difference in mineral content.

Discussion with the investigating officer revealed that the main road bordering the parking lot and field had gravel shoulders on both sides. The officer collected three soil samples from the area between the road and the field. These samples differed from each other in color and texture, but all contained limestone, and one of the soils had the same color and texture as the deposit on the right heel. This supported the interpretation that the deposit on the left heel could have originated from an area where the field and shoulder soils had mixed. Investigators also collected soil samples from two locations along the gravel shoulder on the other side of the main road, where the suspect's bicycle was parked. These soil samples contained limestone with the same color and texture as soil on the soles of both the left and right shoes.

From his examination of the soil on the suspect's running shoes and in the collected samples, Wolfe was able to demonstrate that the soil samples from the left and right heels compared with samples collected from two locations on the side of the road where the body was found. Soil samples from the soles of both shoes compared with samples from the side of the road where the suspect's bicycle was parked. The evidential significance of the shoe deposits aided in the police investigation, which led to the suspect's confession to the crime.

The Servizio Polizia Scientifica, the laboratory of the Italian police in Rome, has one of the largest and finest groups of forensic geologists in the world. They have been successful in educating police and evidence collectors on the importance of gathering soil and related material samples early in an investigation. The Servizio's expertise was tested in a case that began in July 2002, the day before a Jewish religious celebration, when city officers found that Jewish tombs in the Verano monumental cemetery in Rome had been severely damaged. Headstones were broken and some graves had been opened, uncovering coffins. Newspapers and television bulletins reported the event, the first anti-Semitic act to take place in the Rome monumental cemetery. In addition to the inherently disturbing nature of the case, officials were concerned about possible links with neo-Nazi groups in Rome or connections with conflict in the Middle East.

After some investigation, police began to suspect a group of unofficial gardeners who spent time beautifying the cemetery. During a judicial site survey at the cemetery to collect evidence, site surveyors seized the gardeners' equipment, including three picks and two iron bars, from a box inside the cemetery. The picks and bars had white marks and traces of soil on them. The gardeners claimed they had used the picks to restore some partition walls in the graveyard using cement. Investigators collected samples of soil inside and outside the Jewish area, pieces of damaged headstones, and cement from the partition walls so they could compare them with the soil traces and marks found on the tools. Forensic geologists Marco Allievi, Rosa Maria Di Maggio, and Leonardo Nuccetelli began a detailed examination using the stereoscopic microscope, X-ray diffraction, scanning electron microscope analyses, and observation of soil thin sections with a polarizing microscope. The soil analyses showed a full correspondence between the morphological, chemical, and mineralogical features of the sample collected in the cemetery and the ones found on the tools. The soils contained quartz, calcite, plagioclase feldspar, kaolin, and analcime. Since the picks and the bars were found inside the cemetery, the presence of graveyard soil on the tools was unremarkable and provided no new evidence to investigators.

However, the white marks on the tools proved very interesting. Had they been caused by impact with headstones, not cement as the gardeners claimed? The headstones were made from marble, travertine, and limestone marls. Examination of the marks with a stereoscopic microscope revealed that they were doughy and compact, but easy to remove. There were also micrometric stripes on the surface of the tools consistent with up-and-down movement of the tools. X-ray diffraction showed that cement collected from the partition walls contained calcite, portlandite, and larnite. Finally, applied to the marks on the picks and bars, X-ray diffraction showed calcite in abundance and smaller quantities of quartz and plagioclase feldspar. In other words, the headstones contained the same minerals as the marks on the tools, in the same percentages. The scientists used the scanning electron microscope to establish the chemical composition of the material on the tools, the rock samples, and the cement. The marks and the headstone samples contained calcium in abundance and smaller quantities of aluminum and silicon. The cement was different. The physical evidence told the true story; the suspects were lying. Their tools had indeed been in contact with the headstones, and they had done the damage.

In 1955, Carlos Martin Molina-Gallego was the first forensic geologist hired by the laboratory of trace evidence of the National Institute of Legal Medicine and Forensic Science in Bogotá, Colombia. Later, through conferences and educational programs, he began teaching the value and reliability of evidence to judicial police, judges, lawyers, and university

Body of Columbian murder victim Cesar Vargas —COURTESY OF CUERPO TÉCNICO DE INVESTIGACIÓN, FUSAGASUGA, COLOMBIA

students. One of the successful cases that he has worked and testified to occurred in 2007.

The body of Cesar Vargas was found in a clandestine grave on a ranch near Fusagasuga, Cundinamarca Department, Colombia. A team of investigators searched the area and recovered material evidence from a tool shed that was approximately 1,000 feet from the grave. Soil samples were obtained from several tools that included a posthole digger, a hoe, and a shovel. Soil samples obtained from these tools were sent to the National Institute of Legal Medicine and Forensic Science in Bogota. For comparison, samples of the soil from the grave site were also included.

Molina Gallego conducted mineral, chemical, and physical forensic examination and description of the soil samples. The similarities were sufficient to allow him to conclude that the samples from the grave site were consistent in composition with the soil sample taken from the shovel.

Colombian authorities had accused Omar Quintana of committing the murder, but he denied any involvement. However, after the search warrant and the development of the geologic evidence, he was interrogated again and finally pleaded guilty. Quintana told the police that Vargas was a guerrilla member who was extorting him and that is why he was killed.

This accusation could not be verified. Omar Quintana received a prison sentence of eighteen years.

In his book *Science vs. Crime,* Eugene Block tells a remarkable story of geologic evidence involving concrete, slag, the laboratory of the Federal Bureau of Investigation, and the Washington, D.C., Metropolitan Police back in 1958. It is a wonderful example of the interplay of geologic evidence and investigation. Three little boys were fishing in a shallow section of the Anacostia River when one hooked a heavy object. It turned out to be the body of government worker Ruth Reeves. Attached with wire to Reeves's leg was a concrete block. Washington police captain Lawrence A. Hartnett examined Reeves's apartment and found a picture of her with a man whom neighbors identified as "Phil." One neighbor volunteered that Phil was extremely jealous and had threatened to kill Reeves. Phil turned out to be forty-year-old Philmore Clarke, a respected employee of the housing authority. When police visited Clarke's home, they observed that two concrete blocks were missing from the wall surrounding his front garden. In addition, they found a coil of wire in the rear hall that looked like the wire found with the body. Clarke's car was missing, and he claimed it had been stolen. It was found not far from his house, and police believed that Clarke's friends had hidden it.

Forensic geologists at the FBI laboratory examined the concrete block found with the body and other blocks from Clarke's yard. They were identical in composition. Examination of the wire from the body and from Clarke's house produced similar results. At this point, Clarke became a serious suspect, and investigators searched the car for trace evidence. Microscopic examination of debris removed from the car with a vacuum produced small particles of black silica slag. Similar particles were found in the clothing of Clarke and Reeves. Detective Hartnett immediately returned to the place where the body had been found and examined and collected soils from the area. In a sample nearest to where the body was recovered, Hartnett recognized the shiny black slag. Hartnett learned that an electric power plant a few miles away had produced the slag. Plant workers normally disposed of the slag on private property, but one load had been used experimentally to surface a short section of road. That section of road was adjacent to where Ruth Reeves's body had been found. When Clarke appeared before Judge John Sirica, he pleaded guilty and received a penitentiary sentence of five to twenty years.

In 2001, the California Association of Criminalists honored California Department of Justice forensic geologist Marianne Stam for her work, especially for her presentation of information and evidence in a Southern California rape and assault case. The award was well deserved; Stam developed the basis for the evidence through her personal study of the geology of the crime scene. In 1999, a woman and her small children accepted a ride from a male acquaintance. The acquaintance took them to an isolated location on the banks of the New River in Imperial County, California. There he raped the mother, attempted to drown her, and slit her throat. Amazingly, the mother survived. She and her children escaped and hid for thirty hours in the river. Shortly after the woman reported the crime, a suspect was arrested at his residence. He had showered, and his girlfriend had helped him clean his clothes. He did not, however, clean his shoes. He denied any contact with the victims or the crime scene. He said scratches found on his upper left arm and his lower legs were from his girlfriend and his work.

The victims identified the crime scene south of the Salton Sea and northwest of El Centro, California. The woman was able to pinpoint the place on the riverbank where she and the attacker had entered the river. At that place, the New River cuts through different levels of river terraces. Stam visited the crime scene and collected soil samples from the spot on the riverbank and the surrounding area. Most important, she observed the various soil types in the area and their distribution. She noticed a white, windblown deposit composed mostly of quartz and small shells of a freshwater species in the immediate area of the crime scene and in other patches in the area. She also found this material on the suspect's right shoe. Since the shells had wide distribution in the area, they were not significantly valuable as evidence in themselves. However, studies of color, mineral grain shapes, and heavy and light minerals demonstrated a remarkable similarity between the soil from the suspect's shoe and soil from the crime scene. This, combined with Stam's observation of the windblown material's distribution in the area, was compelling evidence that contributed to the conviction of the suspect.

A case that illustrates the importance of chance and the unusual began on September 26, 2002. A report came in of a motorist needing assistance at a small bridge crossing the Shenandoah River in the area of Front Royal, Virginia. Deputies of the Warren County sheriff's office found one man dead from a shotgun blast to the face and another critically wounded by a shotgun. Local law officers knew the victims were involved in the local drug trade. The surviving victim suggested Lewis W. Felts as a possible suspect. Felts was also known to law enforcement for involvement in the drug trade. Lead investigator Sergeant James W. Cornett learned that Felts lived in Alexandria, Virginia, approximately 80 miles away. Cornett contacted the Alexandria police, who placed Felts under surveillance. Soon they informed Sergeant Cornett that Felts was preparing to clean his vehicle, a Jeep. Cornett asked them to take Felts's vehicle into custody immediately to prevent loss of possible trace evidence. This was an important decision since, it turned out, the Jeep had accumulated a large amount of river sand, despite the fact that Felts lived in a city and maintained his vehicle with pride. Could the sand help place the vehicle at the crossing of the Shenandoah River in western Virginia?

A sheriff's detective in another Virginia county, Erich Junger, was asked to examine soil samples collected from the Felts Jeep and from the crime scene. Junger has had a most remarkable career. While serving as a chief pharmacist mate on a nuclear submarine, he taught himself numerous subjects and obtained a doctoral degree in forensic geology. Appointed as a forensic scientist at the U.S. Armed Forces Institute of Pathology, Junger worked many cases for the Department of Defense. When he retired from the Navy, he became a law enforcement officer.

Junger found that samples from the Felts Jeep and the crime scene compared in color, particle-size distribution, and mineralogy, including the heavy minerals spodumene and tourmaline. Most interestingly, both sets of samples contained the green and blue copper minerals malachite and azurite. Approximately one mile upriver from the crime scene was a source of these minerals, an old quarry where copper had been mined years ago. Malachite and azurite are relatively soft on the Mohs scale and do not survive long in active river transport. It was not surprising, therefore, that they were not found in river sediment one mile downriver from the crime scene or upriver from the quarry. This evidence strongly suggested that Felts's Jeep had been at the crime scene and had collected river sand at the crossing. Faced with this evidence, Felts pleaded guilty one day before his trial to one count of capital murder and one count of attempted capital murder. He is currently serving his sentence in a Virginia prison.

Geophysical methods often aid crime investigation. Forensic geophysicist G. Clark Davenport was one of the founding members in 1991 of NecroSearch International. This Denver-based organization uses a multidisciplinary approach to find buried bodies and evidence in cooperation with law enforcement. NecroSearch has worked more than two hundred cases in twenty states and six foreign countries. Geophysical tools and instruments such as metal detectors and ground-penetrating radar are important to their work. Davenport's steady stream of successes include locating a body that had been buried under a concrete slab for more than twenty-eight years and locating a car that contained the body of a murder victim in the Missouri River seven years after the victim's death.

What Davenport calls his most interesting and satisfying case began when NecroSearch was asked to help locate a pickup truck in the Columbia River near Portland. The truck belonged to an ex-convict who had served time for child molestation. After his release from prison, he had found employment with a tire company. He had last been seen fourteen months earlier driving away from a warehouse after picking up a load of tires for his employer. That same morning, a man molested a girl who lived near the ex-convict's home.

The Portland Police Bureau learned that the Coast Guard had reported many tires floating in the river at midmorning on the day the suspect disappeared. The Police Bureau also learned from the tire shop owner that the subject was a hard worker, so trusted that it was he who made the shop's weekly bank deposit. Working with the U.S. Army Corps of Engineers and the Coast Guard, Davenport and his colleagues developed information on the last fourteen months of river flow, sedimentation, scouring, and flooding. Aerial photographs provided detailed information on travel routes from the tire warehouse to the tire shop. Magnetic modeling of the mass of a pickup truck engine showed that it should be detectable by magnetic surveying.

NecroSearch sent two geophysicists to Oregon to perform a magnetic gradiometer survey in the Columbia River between the downstream location where the Coast Guard first noted the tires and an upstream location where the subject may have entered a road bordering the river. The Corps made a work barge available and the Coast Guard provided real-time differential global positioning. The magnetic profiling identified a number of anomalies, some of which were consistent with the modeling of the truck engine. These anomalies were plotted as targets on a river map. The following day, the Police Bureau dive team retrieved a license plate, detached from the first target, to the surface. The license plate was from the suspect's vehicle. Divers also reported a body in the truck's cab, which turned out to be the body of the missing suspect. Investigators used the known times of events on the morning of the disappearance to demonstrate that the suspect could not have molested the child because he was already dead. Medical evidence suggested that while driving along the river, he had perhaps suffered a severe coughing fit from smoking and lost control of the truck, which flipped several times and fell into the river. Evidence in some cases leads to conviction and in others demonstrates innocence. The forensic scientist only provides information and opinion.

An organization similar to NecroSearch exists in Great Britain. The Forensic Search Advisory Group is a consortium of individuals with expertise in finding buried or hidden remains. All members of the group have direct crime-scene experience and are committed to the use and development of technical and field skills for forensic purposes. The group also advances methods of detection and recovery through a program of basic and applied research. The organization has been involved in international investigations. Organizations like NecroSearch and the Forensic Search Advisory Group conduct two types of searches. The first kind is of a specific location where someone has indicated a body might be, but little else is known. In these cases, the problem is to verify the information. In the second type of search, the location of a body is unknown, but a person is missing and there may be a suspect.

The case of the very special brick began in Great Britain on July 9, 1991, when 18-year-old Leeds prostitute Julie Anne Dart was kidnapped. Her kidnapper forced Dart to write a letter to her family informing them that she had been kidnapped and that they were to tell the police. At the same time, the Leeds police in Yorkshire received a letter stating that a prostitute had been kidnapped and demanding a ransom of £140,000 for her safe return. The kidnapper specified that the person with the money should await instructions at a phone kiosk in Birmingham New Street Railway Station on the evening of July 16. On that day, an appointed person answered the phone, but the line went dead. No further calls were made to that phone, and the operation was abandoned.

Three days later, Dart's body was found wrapped in a sheet in a field south of Grantham in Lincolnshire. She had been battered several times with an implement, then strangled. Within days, the police received another letter from the kidnapper expressing regret for Dart's death, but demanding the original ransom sum if further abductions were to be prevented. The murderer sent additional instructions, all of which resulted in no contact.

At about this time, Motorway Maintenance spotted a small cylindrical object with a red light in the southbound lane of the M1 highway south of Leeds near Barnsley. Suspicious of the device, the Military Bomb Disposal

Team closed the motorway and exploded the device. Inspectors later found the device to be harmless. Close by, they found a brick, painted white, with a letter attached.

On January 22, 1992, Stephanie Slater, a 22-year-old estate agent from Great Barr in Birmingham, had an appointment to meet a potential buyer of property. In a crime seemingly unrelated to the Dart murder, Slater was kidnapped and raped. Slater's abductor left a recorded phone message demanding a payment of £175,000. Slater was held for eight days before being released after the ransom was paid. Later that spring, police broadcast the phone message on the BBC's *Crimewatch UK* television program, and one-legged Michael Sams was arrested after his first wife recognized his voice.

Forensic geologist and industrial mineralogist Andrew Smith examined the white painted brick found with the letter on the M1 road. Visual inspection and measurement of the brick's weight showed that it was dense and well-fired, with a red-purple body color. Examination revealed the brick to consist of quartz, cristobalite, mullite, anorthite, ilmenite, hematite, and cordierite. Anorthite, a calcium-rich feldspar, is relatively unusual; most bricks in the United Kingdom contain potassium-rich feldspar. Sodium- and calcium-rich feldspars are usually used in the manufacture of ceramic glass. In addition, cristobalite is typically found only in bricks fired to above 1,100 degrees Celsius (2,000 degrees Fahrenheit).

The chemistry of the brick showed similarities with that of the Etruria Marl Formation, part of the Coal Measures in British sedimentary rocks of Carboniferous age, but it had a higher calcium content. Etruria Marl is normally low in calcium, but two quarries in the Chesterton area of Newcastle-under-Lyme in Staffordshire were known to have calcic Etruria Marl. Letters stamped on the brick identified the brick manufacturer. The brick, of non-standard dimensions, was part of a special order that went to only two brickyards. One was 650 feet from suspect Sams's workshop. The workshop was found to contain bricks that were physically, mineralogically, and chemically identical to the brick found with the letter. Sams, found guilty of kidnapping, murder, and demanding monies with menaces, was sentenced to four life terms by Nottingham Crown Court in June 1992. Later he confessed to the crimes and identified the location where he had buried the ransom money. The money was eventually located using ground-penetrating radar.

A CASE THAT ILLUSTRATES the importance of unusual cultural artifacts in the soil began on a warm East Los Angeles afternoon in 1999 when a male laborer's partially clothed body was discovered lodged between the back seats of his own minivan, parked on a busy residential street. A seasoned homicide detective, aware of the value of physical evidence, observed and collected soil from the vehicle. The detective noticed large clumps of soil near the wheel wells, some of which had fallen onto the street directly below the vehicle.

Police officers apprehended a suspect and searched his residence. A concrete driveway led to his bungalow. Debris littered the front yard next to the driveway. Large, barren areas of exposed, moist soil with tire tracks indicated that vehicles often parked in the front yard off the driveway. The tire impressions were too poorly preserved to help identify any tire manufacturers. Investigators collected soil samples from the driveway and the front and back yards. They also collected several soil samples within a 10-mile radius of the suspect's house to evaluate regional differences in soil color and composition. William Schneck, a skillful forensic geologist with the Washington State Patrol Crime Laboratory Division and owner of Microvision Northwest, examined the samples.

Schneck characterized samples collected from the vehicle and compared them to control soil from the suspect's residence to determine if the soil on the vehicle originated in the suspect's yard. Laboratory examinations conducted by stereo-binocular and polarized-light microscopy revealed similarities in soil color and mineralogy. In addition, Schneck observed man-made materials such as plaster particles, paint chips, and glitter flakes. He found red, green, white, and brown granules—all of a similar size and shape—in both the known and questioned soil. These granules were surface-coated with paint. What could be the source of these particles? Schenck immersed the colored sand grains in epoxy and polished them. He performed optical characterization of the grains in reflected polarized-light microscopy. Further examinations revealed tar and fiberglass adhering to several of the grains. A comparison of these to asphalt roofing shingles of known composition revealed the source. Asphalt shingles naturally decay, and the asphalt releases the rock granules over time. Rain and wind flush the loose granules off the roof, and they intermix with the soil around the building.

At the trial, Schneck explained the similarities in color and mineralogy between the soil from the victim's vehicle and soil from the suspect's yard. He also noted that both samples included similar man-made trace materials, including the roofing shingle grains, fiberglass, and asphalt, allowing

Schneck to express with confidence the probability that the victim's vehicle had been in the muddy area just off the driveway at the suspect's house. In theory, the suspect killed the victim, lodged him in his own vehicle, and drove it approximately 3 miles away, where it was eventually found. The suspect was found guilty of first-degree murder.

In a recent case that does not fit the pattern of most soil evidence but clearly illustrates the contribution being made by forensic geologists, Bill Schneck became involved in the investigation into the serious illness of a small child caused by arsenic poisoning. The suspected person was absolved when Schneck's examination of the child's house revealed a number of mineral specimens left in the house and yard by a former occupant who was a mineral collector. Many of those specimens were arsenopyrite, an iron arsenic sulphide. The child had been eating and chewing on the finely powdered material that had accumulated on baseboards of the house and other places. Lead is not the only material that can cause children problems.

Geoscientists and soil scientists Brad Lee, Tanja Williamson, and Robert Graham worked together to provide evidence in a remarkable case of stolen palm trees. In April 1997, the San Diego County, California, district attorney asked them to study material associated with the theft of $40,000 worth of the trees. The owner had raised the exotic trees from seeds using a particular potting soil, which he purchased in bulk. Investigators collected ten samples of the potting soil from tree roots left in the victim's yard. From the yard of a suspect's home, they collected soil samples from the root balls of thirty-three palm trees, as well as three samples of native soil from the yard. The suspect had seven species of palm trees in his yard. The victim had raised all but one of these species, *Phoenix roebillini,* so those trees could not have been stolen from him. Samples from this seventh species served as controls in the study.

The examiners used several methods in their study, including carbonate determination, color and particle-size analyses, and mineralogical identification. They determined the ratio of light to heavy minerals and identified the heavy ones. In the mineralogical analysis, three hundred grains were counted. Heavy minerals included hornblende, biotite, zircon, epidote, and opaque minerals such as one would expect in potting soil from decomposed granite. Hornblende, the most common heavy mineral, provided the most useful information.

The examiners concluded that twenty-five of the suspect's thirty-three exotic palm trees had been planted in potting soil that compared with that used by the victim. Analyses accurately discriminated the six palms that investigators knew had not been stolen from the victim. There was no evidence to show that the remaining two trees had come from the victim's yard. In a pretrial hearing at which prosecutors presented the soil evidence, the suspect changed his plea from innocent to guilty.

Perhaps the most famous case in which forensic geology provided the key to the investigation was the murder of Enrique Camarena Salazar, an agent with the U.S. Drug Enforcement Agency, and the murder's subsequent cover-up. The case is featured in John McPhee's 1997 book *Irons in the Fire.* Camarena, who was working in Mexico, is believed to have discovered marijuana fields in Chihuahua. The DEA subsequently torched the fields for a $300 million loss to the owners, Mexican drug lords Rafael Caro Quintero and Ernesto Fonseca Carrillo. Agent Camarena was subsequently abducted on a Guadalajara street in broad daylight. During the following month, the White House, the U.S. Department of State, and most law enforcement agencies in Washington put great pressure on the Mexican government and the Mexican Federal Judicial Police—the *Federales*—to find the agent and solve the case. The Americans encountered an attitude summed up in one Mexican law enforcement official's question: "Why are you so concerned about the loss of one agent when we easily lose two hundred drug agents a year?" In addition to a seemingly different cultural attitude toward an agent's disappearance, there was a long history in Mexico of the Federal Judicial Police and other government agencies benefiting from involvement in the drug trade. Now too many people were taking too close a look at that relationship. A decision was made high up within the Mexican police to solve the problem. The Federales went to a farm in the state of Michoacan owned by the Bravo family, known small-time drug runners. The mother, father, and three sons were killed, as well as a Mexican federal police officer. The Federales recovered the bodies of Camarena and his pilot, Alfredo Avelar, on the farm. The recovery of the bodies made news around the world.

This news was of great interest to FBI forensic geologist Ron Rawalt. He suspected that the bodies had been exhumed somewhere else and placed on the farm as part of a cover-up. He based his suspicion on the facts that the bodies were found on the ground wrapped in a sheet; there were no obvious opened graves in the area; and all the potential witnesses were dead. Rawalt sought and received permission to follow up on his suspicions. He immediately called U.S. authorities in Mexico City and asked that any soil adhering to Camarena's body be sampled. A teaspoonful of material, mixed with decomposing material from the body, was the result. Investigators also collected samples from the Bravo farm in Michoacan.

In the laboratory, Rawalt used a plasma-reduction unit to remove the decomposing body material, leaving just rocks and minerals. The samples collected from the farm were dark, greenish-gray obsidian. The samples from the body were white to tan tuffaceous glass. The difference in minerals confirmed that Camarena had not been buried on the farm. Where had he been buried? Rawalt knew that the answer to that question would also prove that the raid on the Bravo farm was a staged cover-up.

Rock fragments collected from Agent Camarena's body were rhyolitic ash. This type of material blasts from volcanoes, is very hot, and travels at great speed. It is the same material that blanketed the Northwest with ash after Mount St. Helens erupted in 1980. Most intriguing was the presence of other minerals in the sample: The black, manganese iron-oxide mineral bixbyite; a beautiful and unusual pink cristobalite; and opal. The cristobalite crystals were elongate, faceted, and clear. The presence of these minerals in cavities in rhyolitic ash is not unusual, but to find them all in one sample is very rare. As FBI forensic geologist Chris Fiedler put it, "Initially you have the whole country of Mexico in which to find where a teaspoon of soil came from." Finding such a combination of minerals put the investigators one step closer to pinpointing the location they sought.

Rawalt went to the library to research geologic papers and maps describing the rocks of Mexico. In Gail Mahood's article "The Geological and Chemical Evolution of a Late-Pleistocene Rhyolitic Center: The Sierra La Primavera, Jalisco, Mexico" in the *Journal of Volcanology and Geothermal Research,* Rawalt found a description of rocks with the same characteristics as those found on Camarena's body.

Rawalt visited with mineralogists at the Smithsonian Institution to review his work. While he was there he was fortunate to meet a visiting geologist who had worked in the Jalisco region. As Rawalt described the rocks, the visitor interrupted to say the mixture only exists in a small part of a Jalisco state park called Bosques de la Primavera. She outlined the area on a map. Rawalt and three other FBI forensic scientists were sent to Mexico to search the area. The Mexican authorities did not welcome them with open arms, and they worked under considerable duress. However, with the help of an informer and, finally, cadaver dogs, Rawalt succeeded in finding a soil sample identical to the earth material from Camarena's body. The search was over. Camarena's original grave had been found, and the cover-up was exposed. Ultimately, in a complex case with international political implications, arrests in the United States alone led to seven convictions for murder and related crimes.

The cases in this chapter illustrate the wide-ranging issues that forensic geologists address in both criminal and civil matters. In following chapters, we will look at the fascinating history of the science and examine in greater detail how minerals, rocks, soils, and related material can provide valuable evidence. We will also look at the methods forensic geologists use; the legal basis for admission of geologic evidence; the field's specific contributions to mine, mineral, gem, and art fraud; and what the future holds for the science.

2
From Sherlock Holmes to the Present

History and legend contain examples of the use of observation, past experience, and evidence in the form of rocks and minerals to solve forensic problems. For instance, in one common old story, shrewd thinkers are able to locate an enemy camp by examining the rocks caught in horses' hooves. In a case in April of 1856, a barrel of silver coins transported on a Prussian railroad arrived with rock and sand and no silver. Professor Ehrenberg in Berlin examined the sand from the barrel as well as samples of sand collected from each stop that the train made. Using his microscope, Ehrenberg was able to identify the railroad stop where the sand substitution occurred. With this information the railroad police quickly identified the railroad employee at that location who was the culprit and he was convicted. However, the formal application of geology and soil science to criminal investigation had to wait for developments in crime-lab technology, as well as the education of investigators about the usefulness of these materials. Since the end of the nineteenth century, geoforensics has been applied widely and has evolved to a high level of sophistication and quality, becoming an accepted and commonly used tool in today's criminal justice system. The use of earth materials as evidence in both criminal and civil matters assists in investigations and serves as evidence in actual court cases.

The idea of professional geologists applying their field to criminalistics began, like applications of many sciences to this area, with the writings of Sir Arthur Conan Doyle. The publication of the Sherlock Holmes series between 1887 and 1927 provided the world with scientific ideas and techniques for solving crimes that until then existed only in the mind of physician-author Conan Doyle. These techniques had never actually been used before. Investigators would only later develop some of the techniques suggested by Sherlock Holmes and use them to solve actual cases.

The venerable Dr. Watson summed up Holmes's knowledge of geology this way: "Practical, but limited. Tells at a glance different soils from each other. After walks has shown me splashes upon his trousers, and told me by their colour and consistence in what part of London he had received them." Today, considering information and techniques available in the late nineteenth century, we know that a real-life Holmes could not really have performed such feats. However, Conan Doyle's writings did plant ideas that now form the basis of forensic geology: The number of kinds of soils is almost unlimited; soils change markedly over short distances; people may collect soil samples on their clothes, tools, or vehicles simply by coming into contact with earth materials; and examination of that soil may help place the person at the location where the soil was collected.

Hans Gross, an Austrian criminal investigator and professor of criminology who may never have heard of Sherlock Holmes, was instrumental in applying scientific methods to the investigation and prosecution of crime. Early in his career Gross served as legal counsel, state's attorney, and later on the appellate court in Graz, Austria. Gross was not only concerned with using scientific methods for solving crimes, but he also studied the underlying causes of crime, the criminal's personality, psychological changes during confinement, and methods of rehabilitation. Spurred by his interest in the rapid developments in scientific method, Gross compiled those methods being applied to crime investigation at the time and published them in 1893 in his classic and practical *Handbuch für Untersuchungsrichter* (*Handbook for Examining Magistrates*). He included what was known at the time about applications in forensic medicine, toxicology, serology, and ballistics. With remarkable foresight and imagination, Gross also suggested many potential applications of science—including earth science—to criminal investigation. Prophetically, he advocated the employment of the microscopist and mineralogist for the study of "dust, dirt on shoes and spots on cloth." He astutely observed, "Dirt on shoes can often tell us more about where the wearer of those shoes had last been than toilsome inquiries." Translated into English with the title *Criminal Investigation,* this book has significantly influenced the development and use of scientific methods in criminal investigation.

The ideas that Conan Doyle published in fiction and that Gross recorded in his forward-looking criminalistics handbook set the stage for the application of geology and soil science to crime investigation. It was only a matter of time until the applications were actually made.

Georg Popp, a chemist by training, maintained a laboratory in Frankfurt, Germany. Like many consulting laboratories in the early twentieth century, Popp's provided chemical and microscopic services in the area of food studies, analyses of mineral waters, bacteriology, and other related

Hans Gross, 1847–1915, generally acknowledged as the founder of scientific criminal investigation. He championed the exacting science of criminal procedure and worked tirelessly to establish the science of criminology in several universities. —COURTESY OF H. LOUIS

Georg Popp, German forensic scientist. In 1904 he developed and presented what many believe was the first well-publicized example of earth materials used as evidence in a criminal case. —COURTESY OF JÜRGEN THORWALD

fields. In 1900, a criminal investigator in Frankfurt who had read Hans Gross's handbook asked Popp to examine spots on a suspect's trousers. From this introduction, Popp's interest in criminalistics developed, and he devoted himself to developing chemical and microscopic techniques for forensic applications.

In October 1904, Popp was asked to examine evidence in a murder case. A seamstress named Eva Disch had been strangled in a bean field with her own scarf. Examination of nasal mucus on a filthy handkerchief left at the scene revealed bits of coal, particles of snuff, and grains of minerals—in particular, the mineral hornblende. A suspect named Karl Laubach was known to work at a coal-burning gasworks and a local gravel pit. In samples from the suspect's fingernails, Popp found coal and mineral grains—particularly, hornblende. It was also determined that Laubach used snuff.

Examination of soil from Laubach's trousers revealed two layers of material. Minerals found in the lower, or deeper, level compared to those in a soil sample from where Eva Disch's body had been found. Encrusted on top of this lower layer was a second soil type. Popp's examination of the minerals in this upper layer revealed a mineralogy and size of particle—in particular, a crushed mica grain—comparable to soil samples from the path between the murder scene and the suspect's home. From these data, investigators concluded that the suspect picked up the lower, earlier soil layer at the scene of the crime, and that mica-rich mud splashed on top of this deeper material as Laubach returned home on the path. Confronted with the soil evidence, Karl Laubach admitted to the crime. Under such headlines as "The Microscope as Detective," Frankfurt newspapers of the day carried articles about the case.

Over a century later, it is impossible to determine how Popp's geologic evidence would stand up in court today. Still, it is remarkable that only a decade after Gross published his landmark book, mineral study was used in an actual case. Gross's prophecy was fulfilled in a real-life example worthy of the fictional Sherlock Holmes.

The case that confirmed the value of geologic information took place in the spring of 1908. Margarethe Filbert was murdered near the town of Rockenhausen in Bavaria. The district attorney in nearby Kaiserslautern, a man by the name of Sohn, wanted to know the source of some hairs found on the victim's hands. He was familiar with Gross's book and had also filed, for future reference, some of the 1904 Frankfurt newspaper articles describing Popp's work in the Disch case. Sohn located Popp in Frankfurt and asked him to study the hair and some other material.

Georg Popp began an intensive study of the material. He determined that the hair came from the victim. Undiscouraged, he extended his studies to other objects, noting with special interest a crust of soil on the dress shoes of the principal suspect, local factory worker and farmer Andreas Schlicher. Schlicher was a person of "low reputation who had previously been suspected of poaching." Investigators established that Schlicher's wife cleaned his dress shoes the night before the murder and that, afterwards, he had worn them only on the day of the murder. However, Schlicher stoutly denied having anything to do with the crime, including walking on that day in the field where the crime occurred. Trousers belonging to Schlicher were found in an abandoned castle near the scene along with a rifle and ammunition. Investigators established that the ammunition was his. However, Schlicher claimed he had left those items at the castle prior to the day of the murder.

Popp collected soil samples from the various scenes associated with the crime and suspect, and studied them with the help of a geologist named Fischer. The suspect's fields contained a distinctive soil containing fragments of porphyry, milky quartz, and mica, as well as root fibers, weathered straw, and leaves. Soil from the scene of the crime contained decomposed red sandstone, angular quartz, ferruginous clay, and little vegetation. Samples from the castle contained coal, abundant brick dust, and broken pieces of cement from the crumbling walls. In addition, Popp observed that the area around the suspect's home was littered with green goose droppings.

Examining the suspect's dress shoes, Popp noted thickly caked soil on the sole of the shoes in front of the heel. The soil must have accumulated on the one day the suspect had worn the shoes—the day of the murder. In addition, Popp reasoned that the layers of soil on the shoes represented a sequential deposit, with the earliest material deposited directly on the leather. Careful removal of the individual layers revealed a distinct sequence. First, directly on the leather, was a layer of goose droppings. Grains of red sandstone lay on top of the goose droppings, followed by a mixture of coal, brick dust, and cement fragments. Popp succeeded in comparing all three layers of material on the shoe with soil from the suspect's home, the scene of the crime, and the castle. Additionally, though Schlicher had claimed he had walked in his own fields on that day, no fragments of porphyry with milky quartz were found on the shoes. Comparison of the soil on the shoes to samples from the scene of the crime and to the location of the discarded trousers indicated that the suspect had lied and had been at those places on the day of the crime.

Although he would go on to make many contributions to forensic science, the Margarethe Filbert case established Georg Popp in the field of forensic geology and set the stage for later studies of soil comparison. In this early case, he had established a time sequence of soil accumulation consistent with the theory of the crime. He demonstrated that two samples from the shoes compared to samples from two locations associated with the crime, and that none of the shoe samples supported the suspect's alibi. Hans Gross had been right. The dirt on the shoes had told more than "toilsome inquiries" could have.

In 1906, Conan Doyle became involved in an actual criminal case in which he applied some of the methods of the fictional Holmes. This was the first of three cases in which the writer became the investigator. An English solicitor was convicted of cutting and mutilating animals, specifically horses and cows. After serving three years in prison, he was released but not pardoned, despite evidence that he was innocent of the crimes. Conan Doyle became interested in the case. In developing additional evidence, he observed that the soil on the shoes the convicted man had worn on the day of the last crime was black mud, not the mixture of yellow sandy clays found in the field where a pony had been killed. This observation combined with other evidence led to a full pardon for the convicted man.

In France, Dr. Edmond Locard, a student of Alexandre Lacassagne, one of the early leaders in forensic medicine, read French translations of *The Adventures of Sherlock Holmes* and parts of Gross's book. It intrigued Locard that the bits of dust a criminal came into contact with during the crime were commonly transferred to the criminal. After he failed a number of times to interest the police in scientific criminalistics, in the summer of 1910 the Sûreté Nationale (national police) finally gave him two attic rooms in the law courts of Lyons and two assistants. This was the beginning of what would later become the laboratory of the Lyons police. The entrance to the laboratory was on a narrow side street around the corner from the great stone building of the Palais de Justice, with its massive Corinthian colonnades and two giant stone staircases leading down to the bank of the Rhone River. Inside the entrance was a gloomy hall with two corridors, one leading to the prison, the other to the dirty, dusty archives. Each day, Locard climbed the winding stairs to the fourth floor, with its coal stove and dismal surroundings. In this setting, he did remarkable work and established the modern approach to the examination of trace evidence.

In 1912, Emile Gourbin, a bank clerk in Lyons, came under suspicion of strangling his girlfriend, Marie Latelle. Gourbin was arrested but produced

Edmond Locard, the French forensic geologist who helped establish the exchange principle, the fundamental tenet of forensic geology —COURTESY OF SKIP PALENIK

what appeared to be an airtight alibi. Locard went to Gourbin's cell and removed scrapings from under his fingernails. The scrapings contained tissue that might have come from Marie's neck, but with the technology available at that time, this was not provable. However, Locard noticed that the tissue was coated with pink dust, which he identified as rice starch. On the dust particles he found bismuth, magnesium stearate, zinc oxide, and a reddish iron-oxide pigment called Venetian red. Examination of the face powder prepared for Latelle by a Lyons druggist revealed a similar composition. Today, with the mass production of face powder, this evidence would have far less significance. In 1912, however, the distinctly special preparation led to the confession of Gourbin.

This case illustrates what is now known as the Locard exchange principle. This principle says that whenever two objects come into contact with each other, there is always a transfer of material. The methods of detection available may not be sensitive enough to demonstrate this, or conditions may cause all evidence of the transfer to vanish after a given time. Nonetheless, the transfer has taken place. This offers the possibility of associating people, objects, and locations involved in a crime with each other. This principle forms the basis for the collection and examination of all trace evidence, from fingerprints to fibers to soil.

The small town of Colma, California, was shocked when, on the night of August 2, 1921, the parish priest, Father Patrick Heslin, was kidnapped. A ransom note was received, but no further contact was made, and the priest was assumed murdered. In a remarkable bit of genius or luck, criminalist Edward Heinrich examined the handwriting on the note and told police that the writer "had the hand" of a baker and decorator of cakes. Thus, when a man named William Hightower reported to police that he had discovered the location of Father Heslin's body, they suspected he might know more than he admitted. Hightower was a master baker.

By 1921, Edward Oscar Heinrich, sometimes known as the "Wizard of Berkeley," was already one of the most famous criminalists in the country. His remarkable work with physical evidence in the areas of paint, fibers, ballistics, poisons, hair, and wood had won the hard-working, skillful California chemist great respect for his ability to use physical evidence to assist investigation and, later, to present the evidence effectively in court. Born in Clintonville, Wisconsin, in 1881, he graduated from the University of California at Berkeley and became a professor of criminology there. Also known for his dramatic style and air of self-assurance, Professor Heinrich made remarkable contributions to forensic science.

Heinrich examined the place on the California beach where Hightower said the body of Father Heslin would be found. In addition to the body, the place produced a number of objects of physical evidence, including boards from the floor of a tent. Heinrich studied grains of sand recovered from Hightower's knife and pronounced them similar to the sand on the beach where the body was found. In Hightower's room, investigators found a tent with sand in it—sand that compared with that from the knife. It appeared that Hightower had kidnapped and murdered Father Heslin, then kept his body in the tent on the beach for several days before burying it in the sand. He then reported receiving information about the body's location to police. At trial Hightower was convicted and sentenced to life imprisonment at San Quentin penitentiary.

In 1925, Heinrich applied his knowledge of geology in an intriguing case. Mrs. Sidney d'Asquith, sometimes known as Mrs. J. J. Loren, had been murdered and her body dismembered. Parts of the body, including an ear, were found in a marsh near El Cerrito, California. An intense search turned up no other parts of her body. On the ear of the victim, Heinrich found grains of sand. Since they did not compare with the black mud of the marsh, he reasoned that the body, with the ear attached, had first been placed elsewhere. Only later were the ear with the sand grains and part of the head removed from the body and taken to the marsh. He studied the sand grains, noting their size and composition. Believing that the salt crystals stuck to them were too few to indicate sand from an ocean beach, he decided the sand came from a river or brook, where it flowed into the sea. He studied maps, looking for the nearest place to the marsh where such conditions existed. The place he suggested was Bay Farm Island, 12 miles from the marsh at El Cerrito and the site of San Leandro Creek. Despite some doubts, authorities organized a search at Bay Farm Island, and the rest of the body was found buried under a drawbridge. The case was never solved. Nevertheless, in this case and others, Edward Heinrich introduced forensic geology to the United States in a most dramatic way.

The Federal Bureau of Investigation Laboratory was one of the first forensic laboratories in the United States to use soil and mineral analysis in criminal cases. As early as 1935, the FBI laboratory was working with soils; by late 1936, in the Matson kidnapping case, the lab used mineral analysis in an attempt to determine where the young victim had been kept prior to his murder. By early 1939, heavy mineral separations and mineral identifications were standard practice in the FBI laboratory in soil cases. The laboratory remains one of the leaders in the world in both research and case studies. Free of charge, the laboratory provides examination and presentation of evidence to any prosecutorial agency, providing that the evidence has not been previously examined by others.

Between 1970 and 1980 in Great Britain, the Home Office Laboratory at Aldermaston, known as the Central Research Establishment, did a remarkable series of studies in forensic geology. Researchers made major advances in the use of color, particle-size analysis, and cathodoluminescence in examination methods, as well as establishing the lack of reliability of some methods, such as the density gradient column. Today, private laboratories and university scientists perform much of the forensic geology in Great Britain because the UK forensic science service has been privatized, is planned for elimination, and no longer examines soils.

In the Chicago area in 1956, Walter McCrone launched McCrone Associates. McCrone's pioneering work with electron and light microscopes, as well as his efforts to teach others about the benefits of their use, was

Walter C. McCrone, 1916–2002, founder of McCrone Research Institute and McCrone Associates —COURTESY OF MCCRONE ASSOCIATES

remarkable. Ultramicroanalysis methods developed by McCrone have been applied extensively in the identification of explosives, particles, and especially contaminants in foods and drugs. In 1960 McCrone established an institute to teach these new identification methods. He established the quarterly journal *The Microscope* and sponsored an annual conference on microscopy. Another contribution was his analysis of materials to establish the authenticity of art objects. Many of today's leaders in the field of forensic geology studied under this remarkable man.

Today, major crime laboratories throughout the world, both public and private, analyze soils and glass. The quality of analysis varies depending on staff training and experience, availability of equipment and time, and the experience of submitting agencies or investigators. Exact figures are not available, but it is safe to estimate that, annually, forensic geologists study thousands of cases throughout the world where geologic material, usually soil and glass, is now routinely collected during investigations.

3 Using Earth Materials as Evidence

Before geologic evidence can be presented in a law court it must first be collected in accordance with the law. That usually means with a warrant, with permission, or incident to an arrest. From the time of collection to the end of the case there must be a detailed and complete record kept of custody, meaning a record of who actually held the samples and the time period over which each person was responsible for the samples. Then the examiner studies the material and draws conclusions about the meaning of the evidence. These conclusions are usually admissible in court only when relevant to the issue in question. The court must be satisfied that the person who examined and is presenting the material is qualified by training and experience to do so. Being allowed by a court to present scientific evidence as an expert witness means being allowed to express an opinion—that a particular bullet was fired from a specific gun, a blood sample came from a specific person, or the like.

There are many kinds of physical evidence. In his classic text *Criminalistics: An Introduction to Forensic Science,* Richard Saferstein describes the following common types:

> 1. *Blood, semen, and saliva.* All suspected blood, semen, or saliva—liquid or dried, animal or human—present in a form to suggest a relation to the offense or persons involved in a crime. This category includes blood or semen dried onto fabrics or other objects, as well as cigarette butts that may contain saliva residues. These substances are subjected to serological and biochemical analysis for determination of identity and possible origin.
>
> 2. *Documents.* Any handwriting and typewriting submitted so that authenticity or source can be determined. Related items include paper, ink, indented writings, obliterations, and burned or charred documents.
>
> 3. *Drugs.* Any substance in violation of laws regulating the sale, manufacture, distribution, and use of drugs.
>
> 4. *Explosives.* Any device containing an explosive charge, as well as all objects removed from the scene of an explosion that are suspected to contain the residues of an explosive.

5. *Fibers.* Any natural or synthetic fiber whose transfer may be useful in establishing a relationship between objects and/or persons.

6. *Fingerprints.* All prints of this nature, latent and visible.

7. *Firearms and ammunition.* Any firearm, as well as discharged or intact ammunition, suspected of being involved in a criminal offense.

8. *Glass.* Any glass particle or fragment that may have been transferred to a person or object involved in a crime. Windowpanes containing holes made by a bullet or other projectile are included in this category.

9. *Hair.* Any animal or human hair present that could link a person with a crime.

10. *Impressions.* This category includes tire markings, shoe prints, depressions in soft soils, and all other forms of tracks. Glove and other fabric impressions, as well as bite marks in skin or foodstuffs, are also included.

11. *Organs and physiological fluids.* Body organs and fluids are submitted for toxicology to detect possible existence of drugs and poisons. This category includes blood to be analyzed for the presence of alcohol and other drugs.

12. *Paint.* Any paint, liquid or dried, that may have been transferred from the surface of one object to another during the commission of a crime. A common example is the transfer of paint from one vehicle to another during an automobile collision.

13. *Petroleum products.* Any petroleum product removed from a suspect or recovered from a crime scene. The most common examples are gasoline residues removed from the scene of an arson, or grease and oil stains whose presence may suggest involvement in a crime.

14. *Plastic bags.* A polyethylene disposable bag such as a garbage bag may be evidential in a homicide or drug case. Examinations are conducted to associate a bag with a similar bag in the possession of a suspect.

15. *Plastic, rubber, and other polymers.* Remnants of these man-made materials recovered at crime scenes may be linked to objects recovered in the possession of a suspected perpetrator.

16. *Powder residues.* Any item suspected of containing firearm discharge residues.

17. *Serial numbers.* This category includes all stolen property submitted to the laboratory for the restoration of erased identification numbers.

18. *Soil and minerals.* All items containing soil or minerals that could link a person or object to a particular location. Common examples are soil imbedded in shoes and safe insulation found on garments.

19. *Tool marks.* This category includes any object suspected of containing the impression of another object that served as a tool in crime. For example, a screwdriver or crowbar could produce tool marks by being impressed into or scraped along a surface of a wall.

20. *Vehicle lights.* Examination of vehicle headlights and taillights is normally conducted to determine whether a light was on or off at the time of impact.

21. *Wood and other vegetative matter.* Any fragments of wood, sawdust, shavings, or vegetative matter discovered on clothing, shoes, or tools that could link a person or object to a crime location.

Of these examples of physical evidence, forensic geology concerns itself with soil, mineralogical substances, and glass, along with any other natural or man-made substances or objects incorporated into soil or other earth materials. Before ever being used in court, of course, physical evidence, including earth materials, may have assisted investigators in numerous ways.

For example, when an exhumed, unembalmed body was discovered one morning in a common green plastic garbage bag on a police pistol range in a New Jersey city, it was assumed that someone was trying to send someone else a message. Only the dirt in the bag that had been dug up with the body tied the body to the original burial site and eventually to those responsible. Detailed examination of the soil led to a map with outlined areas where that type of soil occurs. The soil turned out to have been dredged from a nearby Newark bay for landfill. The landfill could not have been more than a few years old. With this aid to the investigation and assistance from informers, the body's original burial site was located under the front porch of a home built on a landfill. Comparison of soil from that location with samples associated with the body helped tie the body to the original burial site. The body turned out to be the result of a "family affair" in which a mother and daughter murdered the father and buried him underneath the front porch. Distressed by the smell, they dug up the body one night and dropped it in the only wooded area in town—the police pistol range. This example represents the use of soils as both an aid to an investigation and as actual evidence once soil samples from the crime scene are available for comparison.

In another example, in southern Ontario a man was arrested and charged with the beating death of a young girl. The scene of the crime was a construction site adjacent to a newly poured concrete wall. The soil there was sand that had been transported to the scene for construction purposes. As such, the sand had received additional mixing during the moving and construction process and was quite distinctive. The suspect's glove contained sand similar to that from the scene and significantly different in composition and particle size from the area around the suspect's home. This was important because the suspect claimed the soil on the gloves came from his garden.

Larceny of cactuses is big business. Did those giant cactuses in the landscaping of an expensive Southern California home once grow on federal land in Arizona? The only physical evidence tying the specimens to federal land was the dirt attached to the roots. Thieves obtained a bill of sale for hundreds of cactuses from a cooperative private landowner. They then moved onto federal land and dug up the cactuses. Once on the highway, they had the plants and a bill of sale. Who was to say that the cactuses were not removed legally from the private land and were not now being legally transported to Los Angeles for sale to landscapers or private individuals? How do you prove theft of federal property? Without knowing exactly where on federal property the cactuses had been removed, it would be impossible to compare the dirt on the roots with samples from specific locations. However, it would be possible to demonstrate that the dirt could not have come from the land indicated on the private bill of sale. Armed with this information, investigators might then induce one or more of the participants to divulge the true location of the theft so soil samples could be collected from the actual holes from which the cacti were removed, permitting comparison with soil from around the roots of individual plants.

Recently, several similar cases have involved excavation and theft of archeological materials and fossils from federal land. In one case, a pottery shard from an archeological site was physically fitted back together with another shard in the suspect's possession. In several cases, soil encrusted on projectile points, clothing, and pottery found in a suspect's possession compared with soil samples from excavation sites. In a case involving theft of ancient baskets from Manti-La Sal National Forest, Jack Donahue and his colleagues were actually able to collect small samples of soil from under the stitching of the baskets despite the fact that the suspect had cleaned them. In this case, faced with the evidence from the soil analysis, the suspect confessed to the crime. In cases of this type, it is usually critical to establish that the material was removed from federal land and not obtained from private land with permission, as suspects claim.

The examiner may determine that two samples, one from a suspect's shoes and one from the crime scene, compare and have a high probability of having a common source. This opinion may be presented to the court. Alternatively, a suspect may claim that the soil on the shoes was picked up at another specific location. If the soil on the shoes does not compare with soil of the alibi site, then a low degree of credibility may be given to this particular aspect of the story. If the soil does compare with the alibi location, then the case for the alibi is strengthened. The forensic geologist takes the information from the examination and presents the opinion without advocating either side of the legal issue. Some current television programs and books that describe forensic science confuse the roles of evidence collector, forensic examiner, and investigator, giving the public the wrong idea of what forensic science is all about. The true forensic scientist mechanically studies the evidence and presents an opinion independent of advocacy for any side in the legal issue.

In one case that exemplifies these issues, an elderly woman was mugged and murdered in a Washington, D.C., park some years ago. Her body was found under a park bench. Within a short time, a description from a witness who had seen someone leaving the park on the night of the murder led to apprehension of a suspect. It was obvious that the suspect had been involved in a struggle. He had soil on his clothing and in his trouser cuffs. He claimed that he had not been in the park for years, and that the soil came from a fight he had had in another part of the city. Study of the soils near the park bench and from the scene of the alleged fight revealed that the soil from the suspect's clothing compared with that near the park bench and did not compare with samples from the area of the supposed fight. The similarity with the soil sample from the park strongly suggested that the suspect had been in contact with the ground there recently. Furthermore, the lack of similarity between the clothing soil samples and the samples from the area of his alibi cast doubt on his claims.

In another case, soil dislodged from an automobile during a hit-and-run accident provided a clue to where the car had been driven and thus the possible home of the person responsible. The hit-and-run accident, which was fatal, occurred in the upper Midwest. Clumps of soil that dislodged from the fender of the car as it struck the victim and sped away contained the characteristic minerals of the Missouri lead–zinc mining district hundreds of miles to the south. This knowledge contributed to the successful search for a suspect. When the suspect was apprehended and soil collected from under the fenders of his car, it compared with the material from the scene of the crime. It was subsequently shown that the suspect had previously driven through mining areas of Missouri where roads are constructed with material from the mines.

In a substitution case, rocks provided a clue to the place where the switch had been made. When a Canadian liquor store owner opened newly arrived cases of expensive Scotch, he was unpleasantly surprised to find blocks of limestone, each the same weight as a bottle of whiskey, neatly placed in each bottle's compartment. A study of the limestone determined that it could not have come from any of the places through which the whiskey had passed, but that it must have come from its point of origin, Great Britain. Further study revealed that the limestone in the boxes compared with limestone from a particular quarry in central England. Finally, the suspect, who worked for the liquor distributor and had access to the quarry, had often been seen taking home samples of the rock.

When people talk about evidence, they talk about two kinds of evidence or two kinds of characteristics. When we say "individualize" or "having individual characteristics" we mean that the comparison has an extremely high probability of demonstrating that two samples had a common or mutual source. Although we may not be able to demonstrate the probability using conventional statistics, we know from experience and understanding that the possibility of there being another identical example on this earth is essentially zero. This is true when we look at a comparison of fingerprints, random marks on bullets and shell casings, irregular and random wear patterns on footwear and tire impressions, and the fitting together of broken objects in a jigsaw-puzzle-like fashion. But only DNA information can actually produce statistical probabilities that approach individualization.

In certain cases involving earth materials, however, the probability can become so high as to approach that of individual evidence. For example, in a Canadian rape case, the knees of the suspect's trousers contained encrusted soil samples. The sample from the right knee was different from that collected from the left knee. In examining the crime scene, two knee impressions in the soil corresponded to right and left knees. Samples from the two impressions were different. The soil sample collected from the left knee impression compared with that removed from the suspect's left trouser knee, as did samples from the right knee impression and right trouser knee. A major change in soil type occurred between the two knee impressions. Evidence from several cases mentioned earlier, such as the South Dakota homicide of Becky O'Connell and the Montana homicide of Susan Galloway, approach this level of confidence.

Most physical evidence is said to be class type evidence or to have class characteristics. That means that there are probably other examples somewhere on this earth that have the same properties. In the case of earth materials, we can demonstrate a likelihood that two samples have a common origin, even if we cannot always demonstrate a certainty. If we were to take a rock from any outcrop and break it in two, in most cases it would be possible to show, through detailed study, differences between the two pieces. In most cases, however, the similarity between the two pieces would be large. We would be able to say that they compare and that there is a high probability that one piece was a sample of the other, meaning they had a common source. If the two pieces could be fitted together and individual minerals could be shown to line up when fitted back together, then the probability that they were once part of the same rock would be astronomical. In this case, we would say that we had shown an individual characteristic and that there was no doubt about the comparison. We would, however, still be dealing with probability, and the value of the determination would depend largely on the competence of the scientist who made the determination and the availability of data.

In general, recognizing that probability and chance are most important, the usefulness of most types of physical evidence depends on the number of significant variations that can be easily observed or measured in the material. Specifically, how many different classes exist and how widespread is a single class or kind? Because earth materials vary so widely, their potential for use as physical evidence is excellent. If present and properly collected and examined, samples of earth materials rank extremely high in evidential value among types of physical evidence.

Physical evidence, including earth materials, ranks high in our justice system because of the scientific objectivity involved. Many authors writing on the subject of evidence have discussed this. Most people have experienced the sometimes questionable value of statements from eyewitnesses. In contrast to physical evidence, eyewitness testimony is inherently subjective. People may lie; their memories may deceive them; they may misinterpret what they have seen; and different witnesses may perceive events differently.

The following excerpt from an address to the jury by District Attorney Burton R. Laub (Commonwealth v. Lee, No. 58, September Term 1944, Erie County [Pennsylvania]), Court of Quarter Sessions) ingeniously explains the unique helpfulness of physical evidence and some of the principles behind its use. The case included soil evidence presented by FBI Special Agent Richard Flach, who worked on soil for many years at the

FBI laboratory and contributed significantly to the science. This address states the advantages and value of physical evidence.

> Now I appreciate the fact that scientific evidence accompanied by descriptions of such technical instruments as spectrographs and microscopes, and co-mingled with the mystery and magic of test-tubes, melting points, boiling points and other confusing names, means little or nothing to the average layman. I confess that they meant little to me until I started looking into the matter for the purposes of this case. Because of this, I should like, with your permission, to reduce the testimony of these scientists to a simple form so that we can all understand what they mean. Let us take, for example, the testimony concerning paint. Mr. Driscoll told us that he found evidence of five different kinds of paint in the debris which came from the victim's bedclothes and in the debris which came from the defendant's clothing. He told us that these paints existed in the same combination on her bedclothes and his clothing and that, in his opinion, either all of the paint was first on the defendant's clothes and then transferred to the bed or it was on the bed and transferred to his clothes. As another alternative, some of it might have been in both places and then, by contact, became mingled into one mixture of the same elements in both places.
>
> Now we still haven't gotten very far unless we know why he gives us this opinion. You will recall that, on cross-examination, he readily admitted that the types of paint with which we are dealing might exist anywhere and are quite common—although he did say that the black paint in both specimens was of exactly the same chemical composition and that this was a peculiar circumstance since samples of paint from the same bucket are apt to have different chemical compositions. What Mr. Driscoll did say, however, was that though individually these paints might exist anywhere, the probability of their existence in this particular combination was very remote. Now let us see what he means by this. He told us that there was a hard surface red paint, a waxy red paint which he chose to call by another name, there was green paint with an adjacent white layer, blue paint and black paint. For the moment let us forget the word "paint" and talk about something with which we are all familiar. Suppose that I said to you, "I saw a woman today and she was wearing a red hat," and you answered, "I too saw a woman today and she also was wearing a red hat." Now red hats are extremely common; they may be purchased in any millinery store in the country. Therefore, neither you nor I would jump to the conclusion that we had seen the same woman merely because of the color of her hat. But suppose that I said, "My woman was wearing a bunch of waxy-red cherries on her hat," and you responded, "So was my woman." Now, waxy-red cherries are quite common. A few years back they were an accepted decoration for ladies' hats and it would be fair to assume that every attic in the city would

disclose, amid the odds and ends of women's discarded material, at least one bunch of waxy-red cherries. Because of this well-known fact neither you nor I would be willing to venture an opinion that we had seen the same woman. However, we now have developed two points of similarity and are interested in determining whether or not we did see the same person. I describe my woman as having a green cape with a white lining. Garments of this description, while not numerous, may still be found quite commonly; nevertheless, when you reply that your woman was also wearing a green cape with white lining, neither of us have any doubt but that we both saw the same woman. However, we are cautious people and we want more evidence. So you say to me, "My woman was carrying a shiny black pocketbook." Under these circumstances no person of intelligence would conclude that, in a small community such as this, you and I had seen different women. But wait! We have not concluded our comparisons. Suppose that I say, "But my woman was wearing a blue skirt." Now, when you respond that your woman was wearing a blue skirt, both of us will argue to the end of the earth that we have seen the same woman. To clinch matters, however, let us carry our little story a bit farther. Suppose that my woman had dropped her purse on the street and a small chip had fallen off. Because it was so shiny and black, I picked it up. In your case, the woman had bumped her purse against a counter in a near-by department store and you had, for the same reason, picked up a small chip of the black, shiny material which had dropped to the floor. If we take our bits of broken purse to a chemist and he tells us that they are of identical chemical compositions, both you and I will take the witness stand and swear that we saw the same woman. Couple all of these facts with the information that we had seen our woman in the same part of town and at approximately the same time and you will find that we have reduced our probabilities to a certainty.

Now, if we re-translate our colors from clothing back to paint, we have the exact picture as presented here in court. Our red hat is a hard-surface red paint; our cherries are the waxy-red pigment which Mr. Driscoll described. The green cape with the white lining becomes a green paint with an adjacent white layer; the black purse is a shiny black paint and the blue dress becomes, instead, blue paint. That is why Mr. Driscoll had no hesitation in saying that, in his opinion, the two types of debris originated in the same source.

The same type of argument applies with equal force and effect to the expert testimony of Mr. Duggins and Mr. Flach. You will remember how they described the coincidence of brass or bronze particles, cinder and slag material, miscellaneous hair and fiber material, tobacco particles, wood particles, sand grains and salt grains upon the various exhibits. You will recall the colors and textures of the fibers which were found and how they compared. You will remember how, of twenty-four different colors and combinations of fibers, fourteen were present on the defendant's clothes as well as on the bedclothes of the victim.

> It would be too obvious and painstaking to translate these combinations of materials in articles of clothing and draw a similar analogy to the one of the woman with the black pocketbook and red cherries. However, anyone can readily see how this evidence has pyramided beyond the point of speculation to the point where we can say with positive conviction that this is the man who made this vicious attack upon Hilda Miller.

Based on training, experience, professional judgment, and competency, scientists select the measurements and observations that will enable them to make judgments about physical evidence. By the nature of science, these types of observations and measurements should be reproducible by other professionals. There is constant pressure in the crime laboratory to develop standardized, simple, practical methods for comparison. The problem for forensic geology is to apply methods to soil and related samples that allow the scientist to make a professional judgment with the highest degree of confidence. The methods should not be so detailed that they either become so difficult to perform that they are never used or so theoretical that comparison becomes impossible.

Important in this distinction is the recognition that soils, rocks, minerals, and fossils are complex mixtures that nature brought together. In this sense, they are far different from manufactured products such as glass, paint, and fibers. Some scientific methods rely on a single measurement of the total sample, such as a chemical analysis of the entire sample, a measure of the density distribution of the particles in the entire sample, the distribution of the sizes of particles in the entire sample, or the color of the entire sample. Such methods may contribute to the professional judgment of comparison, but they are seldom sufficient in themselves for a professional judgment of comparison. Each method of analysis must also be a standard method that has been demonstrated to be reliable in producing the information it is supposed to produce and has been accepted as reliable.

This is well illustrated in the case of bulk chemical analyses. Such analyses may be excellent if properly performed for the comparison of quality-controlled, manufactured products. However, in the case of soils, it is not uncommon in analyzing two samples from the same spot to have a few mineral grains from one sample that are missing from the other. If these grains have an unusual concentration of a particular chemical element and the number of these grains in the sample area is small, the analyses should show large differences in the concentration of that element. Alternatively, a chemical analysis for certain elements common in minerals may show little difference between a large number of different soils from different places. Typically, unusual minerals and particles are more useful in making a comparison than the more common ones.

In the case of an explosion that destroyed part of a smokeless powder plant, an unusual rock provided useful information for the investigation. In the plant, the powder was extruded from a press through small holes, like spaghetti. The resulting extruded rods were cut to the desired length. The explosion occurred in the press. After the explosion, an experienced investigator found several small rocks on a screen that was part of the press. The rocks must have been mixed into the batch of powder that exploded when squeezed in the press. These foreign objects in the powder were presumed responsible for the explosion. When caught in the press, rocks with a hardness greater than approximately 2.5 on the Mohs scale and a melting temperature greater than 500 degrees Celsius (932 degrees Fahrenheit) produce enough heat through friction to detonate some explosives. The discovered rocks met those specifications. Next, the investigators faced the question of whether the rocks had been placed in the powder deliberately or came to be there by accident. Where had the rocks come from?

Walks, parking areas, and lawns around the plant contained many different kinds of rocks, both local and brought in from elsewhere. Study of rock fragments from the explosion site revealed the feldspar aventurine, sometimes called sunstone, a rare yet distinctive, easily recognized mineral. Careful study of the plant facility identified a place with similar rocks: a sand walk at one of the entrances. This information was used in the investigation, particularly during interrogations of suspects about their movements around the plant prior to the explosion. It turned out that one of the workers had spilled a bag of powder on the sand, then put the powder back in the bag along with some sand. The bag was placed inside the door to be destroyed. However, with no ill intent someone had put the bag back on the production line—with disastrous results.

Forensic geologists work in ways similar to other criminalists. They first search for the unusual rather than the common, knowing that unusual elements or combinations have a high probability of demonstrating a common source. In this search they employ different methods and approaches depending on the kind of soil, the minerals involved, and the size of individual grains. Failure to develop sufficient points of similarity or dissimilarity after intensive study usually results from having an insufficient sample for analysis or failure to observe or measure properties with evidential value. Collection of insufficient sample material or inappropriate sampling commonly results in a meaningless expenditure of effort with little likelihood of a probative contribution.

Forensic geologists examine thousands of samples each year in hundreds of cases. Many of these cases result in conclusions that samples do not compare—they are not judged to be associated. There can be many reasons for this conclusion, including:

- No transfer took place.
- Soil was transferred but later removed by rubbing or washing.
- Two or more soils were transferred, resulting in a composite sample.
- The area has rapid soil changes and the sampling was inadequate.
- The suspect was not at the crime scene.

After performing the study, if the scientist decides on the basis of analysis and professional judgment that certain samples compare or do not compare, he or she must be prepared to defend that judgment as a scientist in a court of law.

The legal basis for which scientific methods are admissible in a court of law is complex. Several court decisions provide guidance for the admissibility of scientific evidence. In Frye (Frye v. United States, 293 Fed. 1013, 1014, D.C. Cir. 1923), the court declared:

> Just when a scientific principle or discovery crosses the line between the experimental and demonstrable stages is difficult to define. Somewhere in this twilight zone the evidential force of the principle must be recognized, and while courts will go a long way in admitting expert testimony deduced from a well-recognized scientific principle or discovery, the thing from which the deduction is made must be sufficiently established to have gained general acceptance in the particular field in which it belongs.

This case involved the question of the admissibility of the polygraph. As science became more complex and specialized, the possibility of finding "general acceptance" in any field by all scientists became a problem. It is uncommon for a specialist in one field to be familiar with methods used by other kinds of specialists. This problem was partly resolved in the Williams case (People v. Williams, 164 Cal. App. 2d Supp. 848,331 P. 2d 251, 1958). In deciding on the admissibility of the Nalline test as a determination of whether a person is under the influence of a narcotic, the court ruled that it only needed to know whether scientific specialists in the field considered the test and its results to be reliable. There remained the question of admissibility of scientific evidence where procedures had been devised to meet a particular problem and where there could have been no opportunity to gain prior general acceptance by other specialists. The Coppolino case (Coppolino v. State, 223 So. 2d 68, Fla. App. 1968, app. dismissed 234 So. 2d 120 Fla. 1969, cert. denied 399 U.S. 927) involved the admissibility of a previously unknown procedure devised by toxicologists to detect a chemical suspected of causing a victim's death. The court decided that such tests devised to explore a given problem were admissible provided the expert witness lays a proper foundation for her or his opinion and explains the accepted principles of analysis used. Within the field of forensic geology, the diversity of geologic evidence involving the identification of rocks, minerals, and fossils; the use of maps showing rocks, soils, and topography; and the application of geologic instruments to forensic problems all made this decision applicable and important.

Federal Rule of Evidence 702 takes a totally different approach to the question of admissibility. This rule states:

> If scientific, technical, or other specialized knowledge will assist the trier of fact to understand the evidence or to determine a fact in issue, a witness qualified as an expert by knowledge, skill, experience, training, or education may testify thereto in the form of an opinion or otherwise.

This rule substitutes the principle of "usefulness" for the "general acceptance" principle of the Frye decision. Some have argued that Federal Rule 702 abrogates the principle of Frye. However, the forensic geologist should certainly be prepared to defend the methodology used and the scientific validity of the results in any testimony, even if the court does not require a strict general acceptance test.

In its 1993 landmark decision in Daubert v. Merrell Dow Pharmaceutical, Inc. (113 S. Ct. 2786), the U.S. Supreme Court said the Frye "general acceptance" standard was not an absolute prerequisite to admissibility of scientific evidence under the federal rules of evidence. In this case, the court decided that newly developed statistical methods should be admitted even if there had not been sufficient time to develop general acceptance. The court ruled that the trial judge would decide whether evidence could be presented in his or her court. This decision applies to all federal courts and is now the rule in many state courts. The Supreme Court provided guidance for judges in deciding admissibility, including:

- Whether the scientific technique or theory can be (and has been) tested.
- Whether the technique or theory has been subject to peer review and publication.

- The technique's potential rate of error.
- Existence and maintenance of standards controlling the technique's operation.
- Whether the scientific theory or method has attracted widespread acceptance within a relevant scientific community.

This ruling has massive significance for the admissibility of scientific evidence. Judges now decide what will be heard in their courtrooms. As with all issues in which discretion is permitted, there are some very positive and some questionable effects. We have already seen a court in New Jersey refuse to admit any evidence not expressed in probability numbers developed through conversional statistics. The New Jersey Supreme Court overturned this decision. A U.S. District Court judge in Pennsylvania ruled that fingerprints were not admissible as scientific evidence, but later reversed himself. It will be interesting to see how the full impact of Daubert plays out.

4
Origin and Distribution of Earth Materials

The value of any material as physical evidence depends in large part on how many kinds there are and how they are distributed around the earth. In this section we will examine the origin, processes, components, and mixtures of earth materials. In so doing, it should become obvious that the number of kinds of earth materials is virtually unlimited, and that they are distributed widely and change rapidly over short distances. For these reasons, the statistical probability that a given sample will have properties similar to any other samples from other locations on earth is very small. The large number of kinds of soils makes the power to discriminate among them equally large, so the evidential value of soil is excellent in many cases. The value increases even more when rare or unusual minerals, rocks, fossils, or manufactured particles are found.

Rocks and Minerals

Over four thousand individual minerals on earth have been identified. Introductory books on mineralogy discuss approximately two hundred of them. Twenty or so are commonly found in soils, but most soil samples contain only three to five minerals.

What is a mineral? Mineralogists generally call something a mineral when a substance is naturally occurring, has a characteristic internal structure determined by a regular arrangement of atoms or ions, and has a chemical composition and physical properties that are either fixed or that vary within a definite range. This definition is difficult to apply to some of the particles we see in soils. For example, both coal and volcanic glass occur naturally. Coal does not have a fixed chemical composition, however, and volcanic glass lacks an orderly internal structure. Many man-made grains, such as silicon carbide abrasive, do not occur naturally, but they do have other mineral-like characteristics.

Geologists can usually recognize on sight twenty to fifty of the most common minerals, as well as several less common ones that possess distinctive properties. They do this in much the same way that a forester identifies trees or a zoologist identifies animals: by observing a characteristic, or series of characteristics, that compares with similar objects they have seen in the past. The characteristics of minerals commonly observed, usually with the aid of a lens or low-power binocular microscope, include color; luster; the way the mineral breaks, called cleavage or fracture; and streak—the color of the mineral when it is finely divided. Some minerals are magnetic, demonstrated by lifting them with a magnet.

The density of minerals varies, another property useful in identification. Geologists generally draw the boundary between heavy and light minerals at 2.89 grams per cubic centimeter. In addition, the form of crystal faces results from the orderly internal structure of atoms and ions. Many different minerals crystallize with the same arrangement of crystal faces. However, when combined with other properties, the crystal form is useful in recognizing some of the mineral species.

In most forensic work, the size of individual mineral grains is small, and the most significant information is often obtained from those minerals that are unusual or uncommon. When geologists cannot identify minerals with a low-power binocular microscope, they use a polarizing or petrographic microscope. This microscope differs from an ordinary biological microscope in that it has filters capable of polarizing light, a rotating stage, and attachments for viewing characteristic effects on light that has passed through minerals.

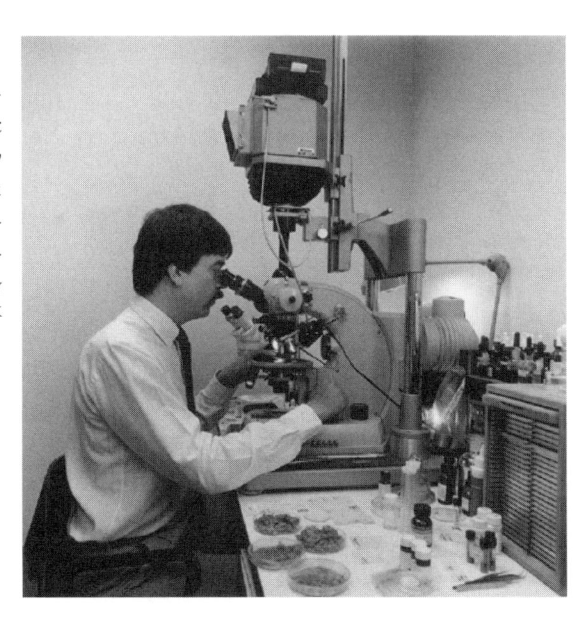

Forensic microscopist Skip Palenik in his laboratory at Microtrace with binocular and petrographic microscopes —COURTESY OF SKIP PALENIK

A common technique in the study of rocks or similar material is to prepare a thin section for study under the petrographic microscope. A geologist prepares a thin section by cementing a rock sample to a glass slide and grinding the sample until it is so thin that it becomes transparent. This permits the scientist to see and identify the minerals and how they fit together.

If minerals of the size of sand and silt are identified, there is generally a predominance of quartz and feldspar with some other minor components. When minerals are found that are generally smaller than 2 micrometers they are commonly composed of a different group of minerals known as the clay minerals. In general the kind of minerals you find in soil depends

COMMON CLAY MINERALS

Mineral	Comments
Kaolinite	Kaolin group clays are the most common raw materials in ceramics. Formed by intense weathering of many silicate rocks.
Halloysite	A kaolin group clay also commonly used in ceramics.
Serpentine group clay minerals (eg: berthierine, odinite, chrysotile [asbestos], antigorite	Common alteration products of basalt and other mafic parent rocks.
Illite	A very common constituent of shales.
Glauconite	Occurs in many marine sedimentary rocks as little green sand-sized grains.
Smectite (montmorillonite nontronite, and beidellite are varieties of smectite)	The most common clay formed by weathering of most silicate parent rocks. Forms in large quantities during weathering of felsic volcanic ash.
Mixed-layer illite/ smectite (I/S).	A very common constituent of shales.
Vermiculite	A common weathering product of micas (biotite and muscovite). Common in soils.
Chlorite	Another common alteration product of mafic rocks. Also a common low-grade metamorphic mineral. Fairly common in soils.
Gibbsite	A common non-silicate clay formed in intense tropical weathering conditions. The major ore of aluminum.
Allophane and Imogolite	Common constituents of rapidly weathered volcanic soils.

—After Graham R. Thompson

COMMON LIGHT MINERALS

Mineral	Approximate Chemical Composition	Crystal System	Specific Gravity	Mohs' Hardness	Common Use
Beryl	$Be_3Al_2(Si_6O_{18})$	Hexagonal	2.75-2.8	7.5-8	Source of Be, gemstone
Calcite	$CaCO_3$	Rhombohedral	2.71	3	Cement, quicklime
Dolomite	$CaMg(CO_3)_2$	Rhombohedral	2.85	3.5-4	Soil conditioner
Feldspar (plagioclase)					
Albite [Ab]	$NA(AlSi_3O_8)$-Ab_{90}-An_{10}	Triclinic	2.62	6	Ceramics, cleaning powder
Oligoclase	Ab_{90}-An_{10}-Ab_{70}-An_{30}	Triclinic	2.65	6	
Andesine	Ab_{70}-An_{30}-Ab_{50}-An_{50}	Triclinic	2.69	6	
Labradorite	Ab_{50}-An_{50}-Ab_{30}-An_{70}	Triclinic	2.71	6	
Bytownite	Ab_{30}-An_{70}-Ab_{10}-An_{90}	Triclinic	2.74	6	
Anorthite [An]	Ab_{10}-An_{90}-$CaAl_2Si_2O_8$	Triclinic	2.76	6	
Feldspar (potassium)					
Orthoclase	$K(AlSi_3O_8)$	Monoclinic	2.57	6	Porcelain, cleaning powder
Microcline	$K(AlSi_3O_8)$	Triclinic	2.54-2.57	6	
Glauconite	$K_2(Mg,Fe)_2Al_6(Si_4O_{10})_3(OH)_{12}$	Monoclinic	2.3	2	Water softener
Gypsum	$CaSO_42H_2O$	Monoclinic	2.32	2	Plaster of Paris
Halite	$NaCl$	Isometric	2.16	2.5	Rock salt, de-icing salt, table salt
Quartz	SiO_2	Rhombohedral	2.65	7	Glass, radio crystal
Talc	$Mg_3(Si_4O_{10})(OH)_2$	Monoclinic	2.7-2.8	1	Optical equipment, soapstone, talcum powder, filler

COMMON HEAVY MINERALS

Mineral	Approximate Chemical Composition	Crystal System	Specific Gravity	Mohs Hardness	Common Use
Actinolite	$Ca_2(Mg,Fe)_5(Si_8O_{22})(OH)_2$	Monoclinic	3.0–3.2	5–6	
Anatase	TiO_2	Tetragonal	3.9	5.5–6	Source of Ti
Andalusite	Al_2SiO_5	Orthorhombic	3.16–3.20	7.5	Refractory
Anhydrite	$CaSO_4$	Orthorhombic	2.89–2.98	3–3.5	Soil conditioner
Apatite	$Ca_5(F,Cl,OH)(PO_4)_3$	Hexagonal	3.15–3.20	5	Gemstone, fertilizer
Aragonite	$CaCO_3$	Orthorhombic	2.95	3.5–4	
Augite	$(Ca,Na)(Mg,Fe,Al)(Si,Al)_2O_6$	Monoclinic	3.2–3.4	1.67–1.73	
Biotite	$K(Mg,Fe)_3(AlSi_3O_{10})(OH)_2$	Monoclinic	2.8–3.2	2.5–3	
Corundum	Al_2O_3	Rhombohedral	4.02	9	Gemstone, abrasive
Diopside	$CaMg(Si_2O_6)$	Monoclinic	3.2–3.3	5-6	
Enstatite	$Mg_2(Si_2O_6)$	Orthorhombic	3.2–3.5	5.5	
Epidote	$Ca_2(Al,Fe)Al_2O(SiO_4)–(Si_2O_7)(OH)$	Monoclinic	3.35–3.45	6-7	
Fluorite	CaF_2	Isometric	3.18	4	Flux for steel, glass
Garnet	A complex silicate	Isometric	3.5–4.3	6.5–7.5	Gemstone abrasive
Hematite	Fe_2O_3	Rhombohedral	5.26	5.5–6.5	Pigments, polishes
Hornblende	$Ca_2Na(Mg,Fe^2)_4(Al,Fe^3,Ti)–Si_8O_{22}(O,OH)_2$	Monoclinic	3.2	5-6	
Hypersthene	$(Mg,Fe)_2(Si_2O_6)$	Orthorhombic	3.4–3.5	5-6	
Ilmenite	$FeTiO_3$	Rhombehedral	4.7	5.5–6	Source of Ti for paint and pigment

COMMON HEAVY MINERALS (continued)

Mineral	*Approximate Chemical Composition*	*Crystal System*	*Specific Gravity*	*Mohs Hardness*	*Common Use*
Kyanite	Al_2SiO_5	Triclinic	3.56–3.66	5–7	Refractory
Magnetite	Fe_3O_4	Isometric	5.18	6	Iron ore
Malachite	$Cu_2CO_3(OH)_2$	Monoclinic	3.9–4.03	3.5–4	
Marcasite	FeS_2	Orthorhombic	4.89	6–6.5	
Muscovite	$KAl_2(AlSi_3O_{10})(OH)_2$	Monoclinic	2.75–3.1	2–2.5	Insulation and filler
Olivine	$(Mg,Fe)_2SiO_4$	Orthorhombic	3.27–4.37	6.5–7	Gemstone
Pyrite	FeS_2	Isometric	5.02	6–6.5	Fool's gold
Rutile	TiO_2	Tetragonal	4.18–4.25	6–6.5	Welding rod coating, source of Ti for paint pigments
Sphene	$CaTiO(SiO_4)$	Monoclinic	3.40–3.55	5–5.5	
Spinel	$MgAl_2O_4$	Isometric	3.6–4.0	8	
Staurolite	$Fe_2Al_9O_6(SiO_4)_4(O,OH)_2$	Orthorhombic	3.65–3.75	7–7.5	Gemstone, ornament
Topaz	$Al_2(SiO_4)(F,OH)_2$	Orthorhombic	3.4–3.6	8	Gemstone
Tourmaline	$(Na,K)(Fe,Mg,Li,Al)_3Al_6(BO_3)_3(Si_6O_{18})(OH)_4$	Rhombohedral	3.0–3.25	7–7.5	Gemstone
Tremolite	$Ca_2Mg_5(Si_8O_{22})(OH)_2$	Monoclinic	3.0–3.3	5–6	
Zircon	$ZrSiO_4$	Tetragonal	4.68	7.5	Gemstone, source of Zr
Zoisite	$Ca_2Al_3(SiO_4)_3(OH)$	Orthorhombic	3.35	6	

in part on the size of the particles. Some minerals simply break up into smaller particles more easily than others. Minerals, especially the small-sized ones, are commonly identified using X-ray diffraction techniques or by other electronic or chemical methods. Also of value in the identification and study of minerals are the transmission electron microscope and scanning electron microscope. With these instruments, scientists can examine particles enlarged over 100,000 times and even identify small mineral particles adhering to larger grains.

Rocks are aggregates of minerals. They may be natural, like granite, for example, or man-made, as with concrete. In nature, rocks form in three main ways: through igneous, metamorphic, and sedimentary processes. The process and material from which the rock is made determine what minerals will be present in it and the texture. Texture refers to the size and shape of minerals and the way they fit together in the rock.

Igneous rocks are formed by the melting of older rocks or parts of older rocks deep within the earth at temperatures commonly greater than 600 degrees Celsius (1,112 degrees Fahrenheit). As the melt cools, minerals grow within the liquid, ultimately producing a solid mass of minerals and sometimes glass. Liquid rock that flows out onto the earth's surface, often from a volcano, is called lava, as is the rock that is formed when it cools.

X-ray diffraction laboratory —COURTESY OF MCCRONE ASSOCIATES

Melted rock within the earth is known as magma. Some igneous rocks form within the earth by the slow cooling of magma. Lavas and magmas have a wide variety of chemical compositions due to the fact that they are formed by the melting of different types of older rocks, or in many cases selective melting of parts of older rocks. This means that there is an almost unlimited number of variations in the kinds of igneous rock.

Volcanic rocks provided the critical evidence in a case discussed by Elisa Bergslien, a noted forensic geoscientist, in her upcoming book *An Introduction to Forensic Geoscience* with Wiley-Blackwell. A Japanese women with the help of her daughter killed her husband and transported the body to the area of Mount Fuji approximately 60 miles from their home. The soil where the body was found contained distinctive volcanic rocks that were black, strongly magnetic, porous basalt with visible crystals of the calcium-rich feldspar bytownite. Similar rocks were found on the floorboard and trunk of the wife's vehicle. The soil around their home was loamy and developed from thick layers of volcanic ash with no basalt fragments. This evidence linked the wife to the crime and led to her conviction.

In addition to chemical composition, the speed at which liquid rock cools is another factor that determines what minerals form. During cooling, different minerals form at different temperatures, and some of the already-formed minerals change to new minerals as the temperature drops. Because the minerals that grow at any given temperature may differ in chemical composition from the liquid around them, the chemical composition of the liquid changes with cooling. For example, if a mineral that needs 10 percent iron forms and the liquid has only 5 percent iron, the liquid will generally have less than 5 percent iron after the mineral has formed. After minerals that form at higher temperatures have grown, a mush of liquid and crystals results. That mush has a different chemical composition from the original liquid. If, as commonly happens, the remaining liquid is drawn off or separated from crystals that formed earlier, the separated liquid acts like new magma and forms minerals that would be expected from magma of the new composition. The melting of minerals and their recrystallization into a wide variety of new minerals produces igneous rocks with an almost unlimited number of possible mineral compositions.

Like the variety of igneous rocks, the textures of igneous rocks—that is, the size, shape, and arrangement of minerals within the rock—depend on the original chemical composition of the liquid rock and how rapidly or slowly it cools. Lava that flows out on the earth's surface and thus cools relatively quickly generally produces small-sized crystals, while magma within the earth, surrounded by rocks that are hot and act as insulators, cools slowly. Slow-cooling magma generally develops larger crystals. In some rocks, minerals that form early in cooling may grow to larger sizes, while the later minerals that form at lower temperatures are smaller. For example, porphyry has larger crystals set in a mass of finer crystals. Such rocks commonly form when magma cools slowly within the earth, allowing the early-forming minerals a chance to grow.

Igneous rocks with the same mineral composition can vary widely in texture. One standard textbook on igneous rock petrography—the description of igneous rocks—lists over 750 different rock types, each with its own range of mineral composition and textures. Anyone who has walked along a stream paved with boulders and cobbles in an area of igneous rocks or looked at a wall made of igneous rocks can appreciate the diversity of igneous rocks.

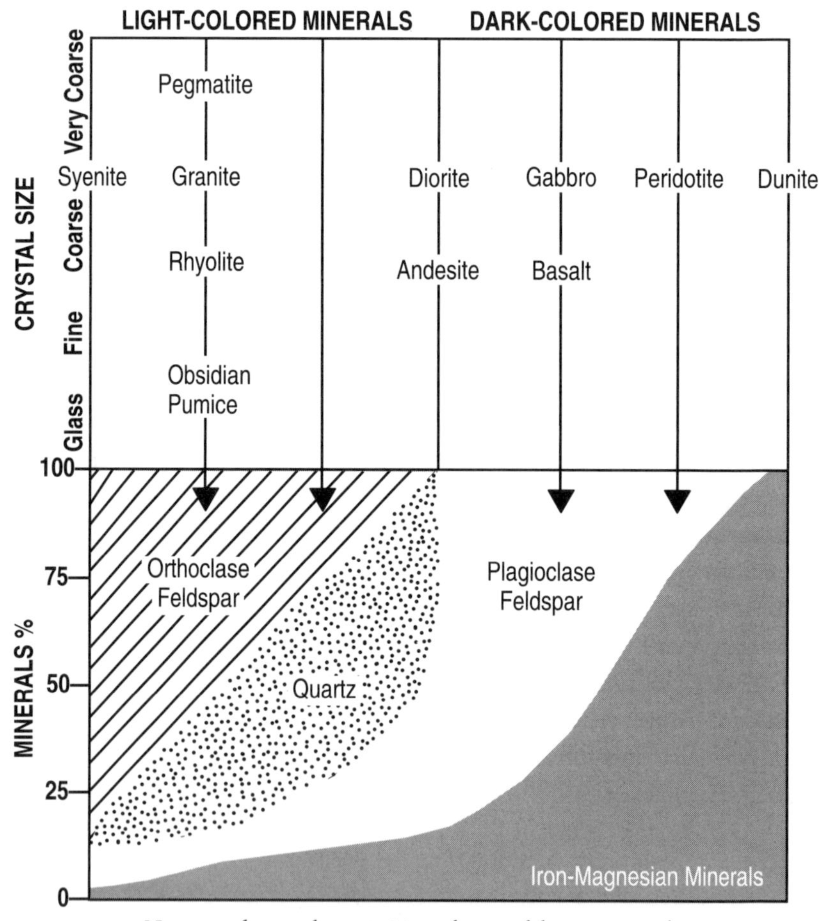

Names and general composition of some of the igneous rocks

Wall made of several types of granite, an igneous rock

Jack Wehrenberg, a most distinguished forensic geologist, who for many years examined the Montana Crime Labs soil cases pro bono, worked a case where a man and a woman were driving on a logging road near Virginia City, Montana. The man was driving and the car went off the road and down an embankment. The woman was killed, and the man was charged with homicide. However, he claimed that he came around a corner and hit a large rock that caused his car to go off the road. Investigators found the rock, a relatively rare igneous rock known as tonalite. The rock showed a scraped and broken edge. Examination of the car revealed that the left rear wheel had fresh fragments of tonalite between the rim and tire. The charge was dropped.

The minerals that make up sedimentary rocks come from the destruction and breakup of older rocks by erosion or weathering. Water, wind, ice, or some other force carries away the broken particles or dissolved elements from the older rock and deposits them at another location on or near the surface of the earth. New minerals then grow in the spaces between the particles, cementing them into solid, sedimentary rock. Generally there are two types of sedimentary rock: chemical and detrital.

Chemical sedimentary rocks are deposited from waters that carry dissolved chemical elements. For example, if seawater is concentrated by evaporation, a variety of minerals will grow and settle to the bottom of the ocean. The most common mineral that forms from seawater is halite—sodium chloride, or common salt. Swim in the ocean on a dry day and let the seawater evaporate as you dry in the sun. Watch the halite crystals grow on your skin. Look closely and you will see several other minerals form. Sea animals and plants also affect the available minerals. For example, simple algae grow minerals within their bodies. When the algae die, the minerals fall to the sea bottom. The same thing happens with the shells of crustaceans or the bones and teeth of fish and other vertebrates. The number of different minerals that have formed from elements from sea, lake, river, or spring water or have formed within the tissues of plants and animals is very large. These minerals make up the particles of most chemical sediments and chemical sedimentary rocks.

The presence of a rock in a place where it could not have originated can constitute substantial evidence in criminal cases. A herd of prize cattle was stolen from a farm in Missouri. Investigators suspected that they had been transported to a ranch in Montana. The brands had been altered, but the herd's owner was certain he recognized his cattle on the suspect's Montana ranch. Examination of the bed of the suspect's cattle truck revealed abundant fragments of chert, a chemical sedimentary rock. This chert was identifiable as originating in Missouri. The suspect denied that his truck had ever been outside the state of Montana. Further study demonstrated that the chert in the truck compared with that at the foot of the cattle-loading ramp on the Missouri farm, where the cattle had gotten it caught in their hooves and had carried it into the truck.

Detrital sediments form from the broken parts of older rocks. During weathering and the breakup of older rocks, new minerals grow from some of the original minerals. Wind, rivers, waves, shore currents, gravity, and glacial ice carry away these new minerals plus the broken minerals from

Sedimentary Rocks: Chemical Rock Types

Rock Name	Composition of Major Minerals	Common Origin
Limestone	Calcite, $CaCO_3$	Shells of marine organisms
Dolomite	Dolomite, $CaMg(CO_3)_2$	Alteration of limestone
Chert	Quartz, SiO_2	Opal shells of marine organisms and chemical precipitation
Gypsum	Gypsum, $CaSO_4 2H_2O$	Evaporation of sea water
Rock salt	Halite, NaCl	Evaporation of sea water
Coal	Altered plant material	Accumulated plant material

the older rock and drop them to form masses of sediment. These masses of sediment take many varied forms. For example, a river may drop sediment on the streambed or along and over its banks during a flood. When rivers reach lakes or seas, they drop much of the sediment they are carrying to form a delta. New Orleans was built on the Mississippi River delta, which formed over thousands of years from sediment deposited by the river. Waves and shore currents carry and deposit sediment to form beaches. Wind blowing over a shoreline can pick up sand from a beach and drop it to make dunes on the inland side of the beach. The same thing takes place wherever plants do not anchor grains of sand or silt and the wind can easily lift them, such as in a desert.

Gravity may start a landslide that moves down a hill, forming a mass of sediment in the valley below. This same process occurs slowly on most slopes, with sediment creeping down the hill under its own weight and forming a mass at the base. Similarly, when glacial ice melts or recedes, it drops the rocks and minerals it is carrying, forming a ridge of sediment called a moraine. If the melt is rapid, the rock and mineral debris will be dropped randomly, littering the landscape with a blanket of sediment.

Identification of detrital sedimentary rocks came in handy some years ago when a gasworks in northern Massachusetts experienced what appeared to be repeated acts of vandalism. The facility's coal supply contained boulders and cobbles of igneous rocks that were destroying the grates in the furnace. The coal was mined in Pennsylvania and transported to Massachusetts by both rail and water. It seemed someone was introducing the rocks into the coal somewhere between the mine and the plant. When examined, the rocks had markings that showed they had been moved by glacial ice. Investigators found that the soil beneath the coal

Sedimentary Rocks: Detrital Rock Types

Rock Name	Major Minerals	Texture
Conglomerate	Rounded fragments of rock	Coarse-grained, over 2 mm
Breccia	Angular fragments of rock	Coarse-grained, over 2 mm
Sandstone	Quartz	Medium-grained, .0625 to 2 mm
Arkose	Quartz, more than 25% feldspar	Medium-grained, .0625 to 2 mm
Siltstone or shale	Quartz and clay minerals	Fine-grained, .0039 to .0625 mm
Mudstone or shale	Quartz and clay minerals	Very fine-grained, less than .0039 mm

storage area near the plant was glacially derived, with similar boulders and cobbles. Further investigation revealed that the coal shovel operator commonly drank alcohol on the job, causing him to misjudge the depth to the bottom of the coal and scoop up underlying glacial deposits. The situation was quickly remedied.

Sediment dropped by glacial ice or gravity sliding tends to contain particles of all sizes. In the case of glacial deposits, boulders the size of a house may be mixed together with the finest clay. Such sediment is said to be "poorly sorted." In contrast, when sediment is moved by a fluid such as water, as in the case of a river, lake, or seashore, or by the wind, the size of particles that can be carried depends on the velocity and viscosity of the transporting medium and the size, shape, and density of the particle. Water, more viscous than air, can move particles more easily. In general, the faster the current, the larger the particle any given fluid can move. If the current slows, larger particles settle out and deposit. When a river enters a lake, it slows, ceasing to flow as a river, and sediment particles settle to the bottom. Larger particles settle out first, and the finest settle more slowly.

But size is not the only factor. A dense mineral will settle faster than a lighter mineral of the same size and shape. Thus, deposits of sediment originally carried by fluids generally tend to be sorted—that is, composed of a limited number of sizes. In the same way, minerals of different densities may be concentrated, as seen in the formation of placer gold deposits. Gold has a very high density, ranging from over fifteen to almost twenty times the density of water. Compared with the common lighter minerals, it is easily dropped when current velocity falls. Because of its density a gold particle settles at the same time as a grain of relatively light quartz many times its size. If the larger grains of quartz and the other common light minerals are not present, the gold will be deposited and concentrated along with the other very heavy minerals. These placer concentrations are the places where gold and other heavy minerals can be easily mined in river deposits.

Sorting and separation by size, shape, and density of minerals are natural processes and result in sediments that differ widely in mineral composition over short distances. When we remember that each river is flowing over rocks that differ widely in mineral composition and texture and that at any place on a river the minerals or combinations of minerals that are possible are all those that might be found upstream, it is no wonder that the possible variations in mineral composition, mineral or rock particle size, and texture are almost unlimited and will change rapidly from place to place.

Detrital sediments deposited by rivers, wind, waves and shore currents, gravity, and glacial ice may turn into solid rock. This takes place when water moving through the holes or spaces between the grains deposits new minerals that precipitate from the water. This precipitate is called mineral cement. Thus sedimentary rock has three parts: the original particles of minerals and rocks; the air- or water-filled spaces between the particles,

SIZE GRADE SCALES IN COMMON USE

Udden-Wentworth	Ø[a] Values	German Scale[b] (after Atterberg)	USDA and Soil Science Soc. America	U.S. Corps Eng., Dept. of Army and Bureau of Reclamation[c]
		(Blockwerk)		Boulders
Cobbles		—200 mm—	Cobbles	—10 in.—
			—80 mm—	
—64 mm—	—6			Cobbles
		Gravel		—3 in.—
Pebbles		(Kies)		
				Gravel
—4 mm—	—2		Gravel	
				—4 mesh—
Granuals				
—2 mm—	—1	—2 mm—	—2 mm—	—10 mesh—
Very coarse sand			Very coarse sand	
—1 mm—	0		—1 mm—	
Coarse sand			Coarse sand	Medium sand
		Sand		
—0.5 mm—	1		—0.5 mm—	—40 mesh—
Medium sand			Medium sand	
—0.25 mm—	2		—0.25—	
Fine sand			Fine sand	Fine sand
—0.125 mm—	3		—0.10 mm—	
Very fine sand			Very fine sand	—200 mesh—
—0.0625 mm—	4	—0.0625 mm—		
			—0.05 mm—	
Silt		Silt	Silt	Fines
—0.0039 mm—	8	—0.002 mm—	—0.002 mm—	
Clay		Clay	Clay	
		(Ton)		

[a] Ø=-log2
[b] Subdivision of land sizes omitted.
[c] Mesh numbers are for U.S. standard sieves: 4 mesh = 4.76 mm, 10 mesh = 2.00 mm, 40 mesh = 0.42 mm, 200 mesh = 0.074 mm.

called pores; and the cement minerals that grow within the pores, binding the particles together. The chemical composition, texture, and amount of the mineral cement add variability to sedimentary rocks.

Upon returning to their New Jersey home, a couple and their teenage son found a burglar inside. Trapped in the living room, the burglar shattered the picture window with a chair, jumped through, and ran down the street. The teenage son pursued the burglar and was steadily gaining on him when the burglar wheeled around and shot him, killing him instantly. Within a few days the police had a suspect. In his room they found shoes with red shale, a detrital sedimentary material, caked on the heels. At the couple's home, outside the picture window, investigators discovered distinctive heel marks in the lawn where the burglar landed. Detailed comparative analysis of scrapings from the suspect's shoes and the soil revealed that the samples from the lawn and the suspect's shoes matched perfectly in every aspect—color, texture, plant parts, and mineralogy. When the suspect was confronted with the evidence, he confessed.

The formation of rocks, a continuing process, has been going on since the beginning of the earth nearly 4.5 billion years ago. Many of the rocks and minerals that become part of new sedimentary rock may have been part of older sedimentary rocks and were not derived directly from igneous rocks. These are sometimes called recycled particles. With this in mind, it is important to recognize that minerals differ in their resistance to wear and also differ in the ease with which they dissolve. A relatively soft mineral such as gypsum, with a hardness of only 2 on the Mohs scale, will be abraded and ground into smaller particles faster than a hard mineral such as quartz given the same exposure to water or wind. Thus we would expect the softer minerals to appear more commonly in finer sizes or to be eliminated during transport. In terms of resistance to dissolving, a given mineral's resistance depends not just on its chemical composition, but also on that of the groundwater or river water dissolving it. Slightly acid groundwater will dissolve the mineral calcite much faster than most other minerals but has little effect on quartz. Thus, what does not dissolve depends not only on the minerals present, but also on all the different climates and waters they may have been exposed to during their history.

In addition, minerals are often subjected to the influence of animals. For example, when an earthworm ingests mineral particles, many physical and chemical changes in its digestive tract cause the destruction of some of the minerals, while others come through relatively unchanged.

It makes sense, then, that sediments subjected to many cycles of weathering and transport contain minerals that are resistant to decay and dissolving,

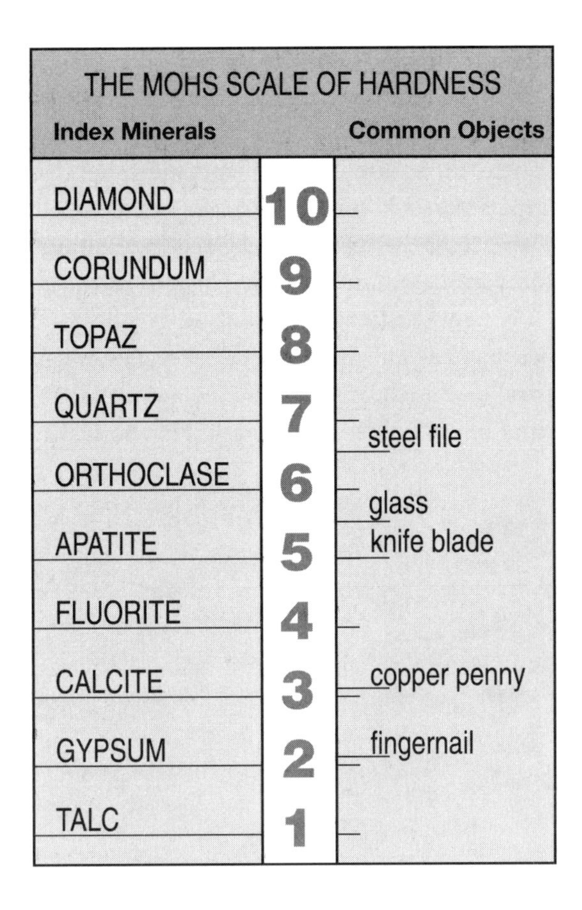

THE MOHS SCALE OF HARDNESS		
Index Minerals		Common Objects
DIAMOND	10	
CORUNDUM	9	
TOPAZ	8	
QUARTZ	7	steel file
ORTHOCLASE	6	glass
APATITE	5	knife blade
FLUORITE	4	
CALCITE	3	copper penny
GYPSUM	2	fingernail
TALC	1	

whereas minerals that are less resistant will be scarce. Most sedimentary rocks contain some fresh minerals and rocks and some material that is old, of which only the most resistant minerals remain.

Igneous and sedimentary rocks may become buried by younger material and carried deep within the earth. This is happening in some areas of the earth that are actually sinking. Southern Louisiana, for example, is sinking at such a rapid rate that sediments deposited by the Mississippi River a few million years ago are now covered by thousands of feet of sediments. At the same time, other areas of the earth are rising. Rocks on top of mountains along the coast of California were deep below sea level and covered by sediments a few million years ago. Buried within the earth's crust, igneous or sedimentary rocks are subjected to very high temperatures and pressures. The temperature within the outer part of the earth increases approximately 1 degree Celsius for every 100 yards below the surface, and the pressure is at least equal to the weight of the overlying rocks. Under these higher pressures and temperatures, new minerals grow within the rocks and the rocks become changed. Such rocks are called metamorphic.

The new minerals that grow to form metamorphic rock occur only under high temperatures and pressures, so they are generally not formed at or near the surface of the earth, but deeper. In addition to these new minerals, textures in metamorphic rock differ from those in the original rocks and are highly distinctive. A shale originally composed of fine detrital particles may become a mass of mica crystals in which the micas are arranged parallel to each other, becoming a mica schist. New minerals such as garnet or staurolite may form within the schist. What forms

Wall made of types of gneiss, a metamorphic rock

Metamorphic Rocks

Rock Name	Texture	Major Minerals	Derived From
Slate	Fine-grained; smooth, slaty cleavage; grains not visible	Clay minerals, chlorite, and minor micas	Shale
Schist	Medium-grained; grains visible; platy minerals parallel to each other	Various platy minerals, such as micas, graphite, and talc, plus quartz and plagioclase feldspar	Shale, basalt
Gneiss	Medium- to coarse-grained; alternating bands of light and dark minerals	Quartz, feldspars, garnet, micas, amphiboles, occasionally pyroxenes	Shale, granite
Quartzite	Medium-grained	Quartz	Sandstone
Marble	Medium- to coarse-grained	Quartz, calcite, dolomite	Limestone or dolomite

depends on the original sediment and the pressure and temperature conditions to which the rock has been subjected. Again we see a process and combination of factors that result in a great diversity of rocks.

The result of all these processes is to produce an almost unlimited number of kinds of rocks. When we look at rocks exposed on the surface of the earth, we are immediately impressed with this diversity. That explains the value of soil evidence: If two samples have the same properties, there is a good probability they came from the same place.

Fossils

Rocks, particularly the sedimentary rocks, often contain the fossilized remains of animals and plants. These fossils can be used for a variety of purposes. The geologist uses them to determine the age of the rock. This is possible because, through evolution, plants and animals have changed drastically through time. Thus we find certain fossils preserved in rocks of one geologic age and absent in rocks of earlier or later ages. Some fossils are found in rocks that were deposited over a longer period of time because the evolution of those plants or animals took place slowly. In other cases a particular fossil or group of fossils survived only a short span of time and will be found in rocks that were formed during that specific time.

The oldest fossils have been found in rocks as old as 3.5 billion years. However, most of these early life forms did not produce hard, decay-resistant body parts such as shells, so few are preserved. Approximately 550 to 650 million years ago, animals had evolved to the point where many were producing hard parts. In sedimentary rocks formed from that time to the present, these parts are sometimes preserved and can be recovered.

Just as living animals and plants today select certain places in which to live, this has also been true in the past. For example, for habitat, a certain species of oyster may require seawater within a narrow range of temperature and with just the right amount of salt, as well as a sea bottom with hard places to which they can attach themselves and grow. Similarly, ancient animals and plants restricted themselves to particular environments. Thus two fossils that were living at the same time may be found in different environments. For these reasons we find a great diversity of fossils in rocks. Over 1,090,000 different species of animal fossils have been recognized.

Some fossils are relatively common and some are very scarce. The people trained to study and identify them, much the same way a biologist studies and identifies living animals, are called paleontologists.

GEOLOGIC TIME TABLE

	SYSTEMS	SERIES	MYA*	EVOLUTIONAL EVENTS
CENOZOIC	QUATERNARY	Pleistocene		Man appears
	TERTIARY	Pliocene	3	Many elephants, horses, large carnivores Mammals diversify
		Miocene		
		Oligocene	22	Grasses become abundant Grazing animals spread
		Eocene		First horses appear
		Paleocene		Mammals develop, expand Dinosaurs die off
MESOZOIC	CRETACEOUS		62	Flowering plants appear
	JURASSIC		130	Many dinosaurs Birds appear
	TRIASSIC		180	Primitive mammals appear Many conifers and cycads Dinosaurs appear
PALEOZOIC	PERMIAN		230	Reptiles spread, conifers develop
	PENNSYLVANIAN		280	First primitive reptiles, abundant insects Many coal-forming forests
	MISSISSIPPIAN			Fishes diversify Amphibians appear
	DEVONIAN		340	Forests appear
	SILURIAN		400	First land plants and animals
	ORDOVICIAN		450	First fish appear
	CAMBRIAN		500	Abundant marine invertebrates
PRECAMBRIAN			570 4,500	Simple marine plants evolve Beginnings of the earth

*Millions of years ago

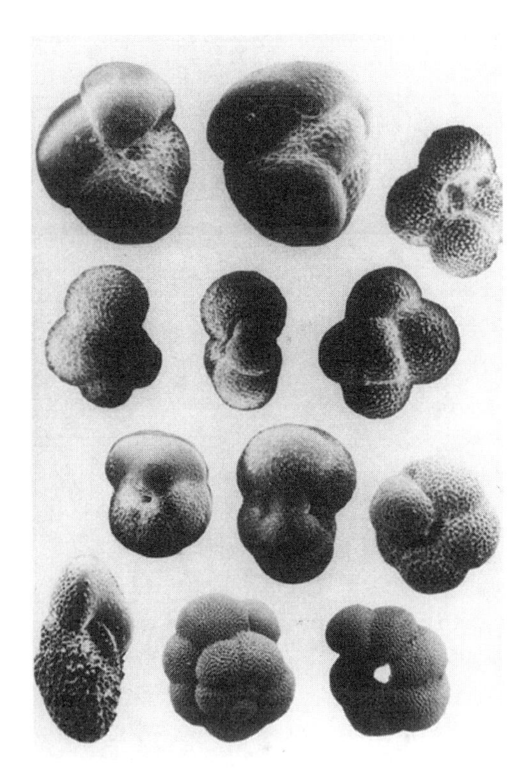

Some variations found in foraminifera, among the many kinds of microscopic fossils —COURTESY OF R. K. OLSSON

Most paleontologists are specialists and become proficient at recognizing on sight one group of fossils—for example, dinosaur bones or corals or microscopic creatures such as diatoms. Sometimes, to identify a fossil, paleontologists compare them to previously identified samples. Fossils add another kind of diversity to rocks, further facilitating the comparison of samples or the determination of a sample's original source. This is especially true for the tiny fossils that can only be seen with the aid of a microscope.

The burglary of a trucking company in Alabama provided an excellent example of the value of fossils as part of soil evidence. Tom Hopen, an examiner with the Alabama Department of Forensic Sciences, developed a most interesting case. The land on which the trucking company was located was quite swampy and a large amount of fill was brought in from another county before the structures were built. The fill material contained many foraminifera microfossils. Hopen determined that the soil on the shoe of the suspect contained the same minerals and the same microfossils as the fill material. This evidence strongly suggested that the suspect had

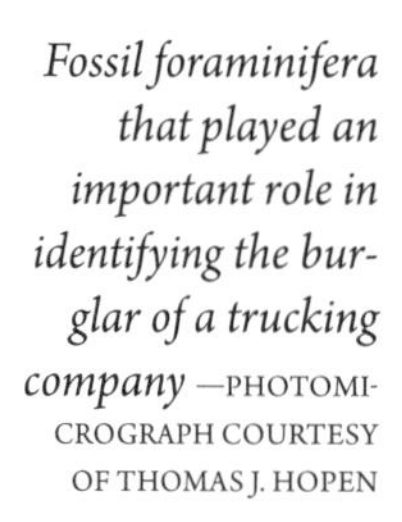

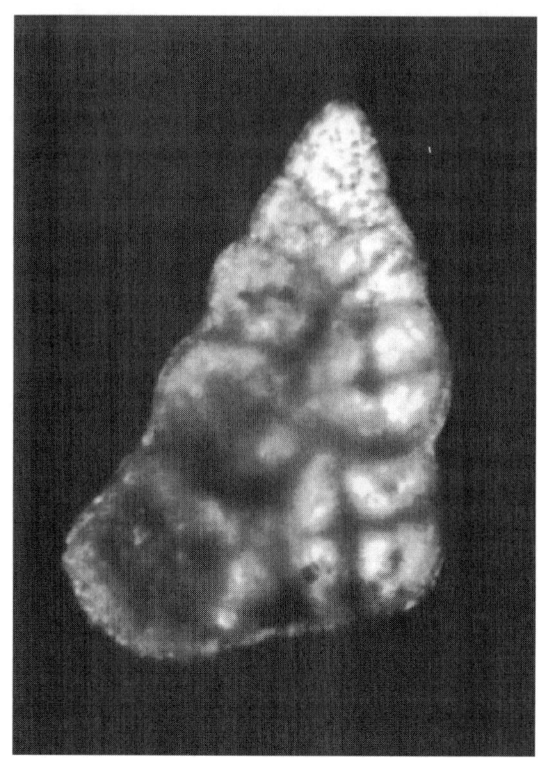

Fossil foraminifera that played an important role in identifying the burglar of a trucking company —PHOTOMICROGRAPH COURTESY OF THOMAS J. HOPEN

walked at the trucking company location or a quarry in another county that was the source of the fill material. Hopen is now a forensic geologist with the laboratory of the U.S. Bureau of Alcohol, Tobacco, Firearms, and Explosives.

The spectacular fossil fish from the Eocene Green River Formation in Utah, Colorado, and Wyoming can be seen in many collections and are sold in mineral and art stores. These fish, very well preserved, are found in the fine layering of the sedimentary rock. The layering is similar to varves found in glacial lakes, where the layers represent a single year. It may be that the lower levels of the lakes were toxic. No organisms could live on the bottom and scavenge dead fish or disrupt the sediment, resulting in beautifully layered sediment and well-preserved fossils. The Green River fish could also have lived and died in desert playas that dried up periodically.

Because many fossils have great monetary value, it is no surprise that people secretly unearth them from U.S. government property or private land and sell them as having been collected legally. A few years ago, FBI forensic geologist Bruce Hall was called in to examine fossil fish from the Green River Formation. A man had been arrested and charged with removing the fossils from federal land. The suspect claimed he had collected the fish on the other side of the mountain with proper permission from the landowner. Fortunately drill cores of the formation existed for study. Hall measured the thickness of each of the annual layers of rock found with the fossils. How thick each annual layer is depends on the conditions during that year. Because each year is different, the pattern of thickness of layers can be matched from place to place. The pattern in the suspect's fossils matched a pattern in the cores that corresponded to the surface level on the government land. On the other side of the mountain, where the suspect claimed he had collected the fish, the matching pattern would have been several hundred feet below the surface. Hall's evidence showed that the fish came from the same level as the federal land and could not have come from the alibi location. The evidence contributed to the conviction of the fossil fish thief.

Coal

Coal is a rock composed for the most part of the fossil remains of plants. Similar material is forming even today in swamps. Water carries detrital mineral grains into swamps, where they mix with the plant material, adding impurities to the coal. As the swamp sediment becomes buried, it gradually changes from the original sedimentary rock into a metamorphic rock according to the following stages or ranks: plant material → peat → lignite → bituminous coal → anthracite coal. The organic material itself contains mainly carbon, hydrogen, and oxygen, with small amounts of nitrogen, sulfur, and other trace elements. As the coal metamorphoses and rises in rank, the carbon content increases and the oxygen and hydrogen decrease, resulting in almost unlimited variation in kinds of coal.

Because coal is commonly used as fuel, coal fragments sometimes turn up in sweepings from automobile floorboards, in the soils of older cities, and in many other locations far from coal mines or outcrops. The presence of coal in a sample, the amount of it, and the kind as determined by microscopic studies have proven useful to the forensic geologist. In addition to the microscopic work, since coals have received considerable study as fossil fuels, differential thermal analyses and detailed chemical composition techniques exist to help distinguish among coal samples.

How Natural Soils Form

For the most part, the earth is made of solid rock. The continents are composed of rocks that are generally light-colored and have a relatively high silica content and, on average, a lower density than rocks found beneath the world's oceans. This density helps explain why the oceans are lower and thus filled with water. The mud and sands covering the ocean bottoms contain a variety of minerals. Underneath this sediment are rocks made from older sediment whose mineral grains have bound together naturally. Beneath these rocks, volcanic lava forms most of the solid foundation of the ocean basins. This volcanic rock is generally dark, relatively low in silica content, and slightly denser than the rocks we see on the continents.

The solid rocks of the continents are usually covered with soil. We glimpse them where soil has been removed, such as in cuts made for highways or railroads, along the banks of rivers, or in sea cliffs or mountainsides. Places where solid rock is exposed at the earth's surface are called outcrops. Most solid rocks are covered with unconsolidated, loose material, composed for the most part of broken up and dissolving fragments of minerals and rocks. For the moment, we can call this soil. The thickness of the soil material—that is, the distance from the surface of the earth down to solid rock—varies from place to place, from a fraction of a millimeter to hundreds of meters.

An investigation of soil helped solve a case in Sebastopol, California, where a group of citizens was opposed to a new housing development known as Laguna Vista. One of the leaders of the citizen group, while walking across the proposed subdivision, found four little green plants that he recognized as the federally endangered plant species, the Sebastopol meadowfoam. He immediately contacted the California Department of Fish and Game which has jurisdiction in such matters. Examination of the root areas of the plants demonstrated that the soil around the roots was different from the surrounding soil. The meadowfoam had been transplanted to the proposed subdivision, a serious federal offense that was obviously done for devious purposes. Following a detailed investigation it was determined that there was insufficient evidence to make an arrest.

Soil material forms naturally in one of two ways: It is either residual or transported. Residual soil material is material formed in place. Solid rock exposed in outcrops at the earth's surface undergoes the natural processes of weathering. These break up and dissolve the rock, turning it into a mass of fragments and removing some of the material, as through the dissolving of minerals in rainwater and groundwater.

The natural weathering or breaking up of solid rock occurs in a variety of ways. Some especially effective mechanical methods are frost-wedging, root-wedging, growth of new minerals, and expansion of minerals. In frost-wedging, water seeps into cracks in the rock and freezes. Ice has greater volume than water and forces the cracks open, fragmenting the rock. Tree roots have a similar effect. We are familiar with both of these effects in the form of cracked and shattered concrete paths and driveways. The growth of new minerals also cracks rock. New minerals may crystallize from groundwater, especially where chemicals in the water become concentrated by evaporation. Some minerals—for example, clays—expand when wet. Expansion can cause rock to break and crumble.

Chemical processes also affect rock. Rainwater and groundwater dissolve some minerals more rapidly than others, leaving holes in solid rock. The holes lead to crumbling and the formation of soil. Long exposure to water can cause feldspars and iron- and magnesium-rich silicate minerals to convert into clay minerals. As such, the mineral grains expand in size, breaking the rock into a loose mass of soil material. Wind or water may carry away some fragments, or they may dissolve, but the remaining soil is produced directly from underlying rock.

In the second method of soil formation, soil material is transported from other areas. Fragments of minerals produced in one place are transported to a new location. Nature provides many ways for transporting soil-producing materials. Fragments of rocks and minerals created by weathering may be carried away by rivers and deposited as sandbars, gravel, or accumulations of fine mud. Wind moves vast quantities of fragments, depositing them as sand dunes or dust layers. Waves and currents along the shores of seas and lakes break up rock and mineral fragments, transport them, and ultimately deposit the particles as beaches or as sediment beneath lakes or seas. Chemical elements carried by water may supply the calcium that shellfish use to produce their shells, which in turn provide special material for soil formation. Gravity causes landslides that move tons of rock and mineral debris down slopes, producing a mass of newly transported soil on the land below.

Of great importance in the northern latitudes and on the tops and slopes of higher mountains is the fact that, over the million or so years ending about 10,000 years ago, our planet experienced periodic changes in climate marked by unusually cold, wet periods known as Ice Ages. During these times, glacial ice formed in northern areas and the high mountains. In the present-day United States, this ice spread as far south as central New Jersey, Ohio, Illinois, and Kansas and covered many peaks in the Rocky

Loosely consolidated river-deposited sands and gravel in three distinct layers, and material from higher layers that has fallen and washed down to a lower level

Mountains as far south as New Mexico. The same was true in northern Europe and Asia and at higher elevations in the Alps. As it moves, glacial ice has tremendous power for grinding up and moving rocks. Picked up by the ice, rocks gouge and grind away the underlying solid rock. When the ice melts, rock debris within the ice drops to the newly exposed ground. This rock and mineral debris is called till or glacial drift. In addition, meltwater streams and rivers carry tremendous amounts of rock and mineral fragments—glacial outwash—away as the ice melts. Winds blowing over the glacier pick up masses of dust and deposit it over wide areas, producing a soil called loess. Much of the soil material in glaciated regions formed as a direct result of glaciation.

A major substitution case worked by Canadian forensic geologist and police service constable Richard Munroe illustrates the importance to forensic geology of glaciation's role in soil development. In 1997, a gold shipment worth $3 million in U.S. currency was moved from the interior placer mines of central Ghana to the coast. The shipment consisted of a series of wooden crates containing placer gold in canvas bags. The crates were flown to London for processing, where they sat in storage for several days without customs inspection. A conflict arose over processing costs and taxes in England, and it was decided to do the work in Canada instead. The crates were moved to Amsterdam and stored for a period of time, again without customs inspection, then flown to Toronto, Canada. There Canadian Customs tagged the crates and gave them additional seals, but again did not inspect them. Finally, an armored car took them to a secure storage facility and later, the processing company.

When the crates were finally opened, the gold had been replaced by sand and pig-iron ingots. Who made the switch and where? It became clear that the only secure handling took place upon arrival in Canada; the Canadian seals were still intact when the crates were opened. However, investigators still suspected that Canadians might be involved. In addition to the crime, there was the question of liability. Three different airlines had been used as well as multiple land carriers and storage facilities. Who was responsible for paying for the missing gold?

Munroe studied the sand using optical and scanning electron microscopes and cathodoluminescence. He noted that the sand was not of glacial origin: It lacked freshly ground minerals and it had been deeply weathered in a tropical or subtropical climate. That ruled out Great Britain and the Netherlands and focused the investigation on the port where the gold was shipped from Ghana. In the sand were fragments of volcanic and sedimentary rocks that showed evidence of having been subject to metamorphic processes. These fragments were consistent with Ghana's geology.

Ghanaian police and government officials were contacted about obtaining reference samples for comparison. They said that, due to rebel activity, accessing the mining district would require an armed expedition. The road system was poor, and much of the journey would have to be on foot. In addition, qualified persons would have to be hired to collect the samples. The Ghanaian police could offer Canadian investigators little help. Nor could investigators contact the gold-producing company for comparison samples, since it might be involved in the crime. In the end, specific studies to discover the actual transfer site were impossible. However, the mining company withdrew its insurance claim. This removed any suspicion from Canada or Canadians and ended the Canadian interest in the case. Munroe's study served the government and people of Canada very well even though what actually happened in Ghana remains a mystery.

Soil is not only the loose mineral and rock fragments we have discussed so far. It is also the mixture of those minerals with organic material, forming the upper part of the loose earth material that supports plants. Soil forms and develops at the earth's surface as a result of the interactions between living and nonliving material. Green plants store solar energy by photosynthesis. When plants die, their residues enter the soil and decay. Decomposition releases energy, and new materials are produced. Among the other processes taking place in soil, many minerals undergo change and even destruction. For the most part, however, such changes take place slowly. Still, soil is a dynamic system in which biologic and chemical processes are constantly taking place. In large part, changes in temperature and moisture influence and control the rate of biologic reaction.

Soil is a three-dimensional body. Soil properties tend to change with depth. Accordingly, a sample taken from a depth of 3 inches may have entirely different characteristics from one taken at a depth of 12 or 15 inches. Thus, when comparative analysis is made between soil adhering to shoes, clothing, or tires and soil from a specific site in nature, investigators must consider not just soil location but also the depth from which the sample came. For example, natural soil from a 3-inch depth may be gray in color with 4 percent organic matter. At 15 inches, it may be yellowish brown with 1 percent organic matter. The various layers, or horizons, of soil are referred to collectively as the soil profile. Soil horizons have their own individual sets of characteristics with respect to appearance, color, texture, and chemical and mineralogical properties. Differences among horizons can usually be recognized visually, then further characterized by laboratory analyses.

How many varieties of soil exist? There is no specific number of soils. Most soil scientists would agree that no two points on the surface of the globe have precisely the same soil. In 1675 in England, John Evelyn, reflecting philosophical discussions of his day, stated that there were at least 179,001,060 different "sorts of earths." In 1909, the U.S. Department of Agriculture listed 230 soil series in the United States, further divided into soil types. For example, Hagerstown (a deep, well-drained soil from the limestone valleys of the eastern United States) is the name of a soil series, whereas Hagerstown clay loam and Hagerstown silt loam are soil types. In 1930 in the United States, there were approximately 1,500 recognized soil series. By 1965 the list had grown to 10,466. Thus the number of recognized soils in any one area often corresponds to the purpose, need, and intensity of the survey. If all the soils in the United States were mapped on a scale of 4 inches per mile—a scale now generally used by the U.S. Department of Agriculture—some 20,000 varieties might be recognized. Even with this large number, further variations would exist within mapped soil units.

The complexity of the processes that form soil is important to the forensic geologist. Considering, as discussed above, that residual soil material develops from so many types of solid rock and is modified by many different climates; that changes in soil composition occur; and that soil particles move and deposit selectively around the earth, it is not surprising that samples of earth materials vary vastly between places.

A case of stolen tobacco leaves, later sold to a warehouse, illustrates the diversity of soil found in a tract of agricultural land. The tobacco leaves had enough soil on them to permit a study. Investigators also took soil samples from the fields of the original owner and the suspects. Investigators determined that the stolen tobacco had grown in the southern half of one of the original owner's 10-acre fields. In addition, the material from the leaves did not compare with the samples from the suspects' fields, disproving their claim that the tobacco was their own.

In the well-known Coors kidnap and murder case, Adolph Coors III, grandson of the brewery founder, disappeared one morning near Morrison, Colorado, southwest of Denver in the foothills of the Rocky Mountains. His automobile was found with the motor still running, and his glasses and splotches of blood were found at the scene. However, his fate and whereabouts were unknown. A month later, a vehicle belonging to a suspect, Joseph Corbett, was found burning on a dump in Atlantic City, New Jersey. Corbett was a suspect because a local miner had remembered part of the license number of a 1951 yellow Mercury sedan that had appeared several times near the Coors ranch. Soil samples from the vehicle's fender showed four layers of earth materials. The outermost, last-deposited layer compared with soil samples from the entrance to the dump, but the interior three layers contained mineral grains characteristic of the Rocky Mountain Front west of Denver. Between Colorado and New Jersey, the car had apparently touched only pavement, acquiring no further soil layers. To help locate the victim, investigators collected over 360 soil samples from the Rocky Mountain Front and began comparing them with samples from the burned automobile. Before they finished, hunters found the body of the victim 27 miles south of Denver. Additional study revealed that the second youngest layer of soil from the suspect automobile compared with the scene where the victim was found; the third layer compared with soil samples from near the bridge at the victims' ranch where his car and blood were found. The fourth and oldest layer did not compare with any of the 421 soil samples ultimately collected and studied in the case, but it probably came from the Denver area. The sample associated with the location of the kidnapping and the one associated with the place where the victim's body was found placed the suspect at the scene. This evidence contributed to the conviction of Joseph Corbett Jr. in a Denver courtroom.

Pollen and Spores

One area of earth science that is seeing increased use in forensic work is the study of pollen and related particles. These microscopic objects, which cause so many problems for hay fever sufferers, have excellent evidential value because so many different kinds of plants exist. Plants produce large volumes of pollen, and the particles reflect the plant communities of a certain area. Those who study pollen and spores are called palynologists. When palynologists work with modern plants, their methods are similar to those of botanists, while their work with ancient plants is closer to that of paleontologists.

In the forensic study of pollen and spores, it is important to know what is produced in a given area and the dispersal pattern there. In addition to the large number of pollen kinds, pollens vary in size and weight according to species. This strongly affects dispersal patterns and what accumulates at a given place. For example, pollen from such plants as marijuana, alder, and birch are very small and light. They fall from the air at speeds of around 2 centimeters per second. Larger, heavier pollen from maize and fir, for example, falls at around 30 centimeters per second. Pollen also appears at specific times of year, adding to its usefulness as evidence. Noted Swiss criminalist Max Frei-Sulzer demonstrated this property in a case in which the suspect claimed his gun had been sealed in a box for over a year and had not been removed and used in a crime. Frei-Sulzer found alder and birch pollen stuck to the oil on the weapon and testified that these trees pollinate at the time the crime was committed.

Examples of common pollen grains —COURTESY OF PATRICIA WILTSHIRE

In a more recent case in Great Britain, a man was accused of rape. The location of the crime was a public garden with a large variety of plants and flowers from various parts of the world. The suspect claimed he had never visited that garden. However, he had difficulty explaining how his jacket had acquired a collection of pollen that contained most of the species found in that unique location.

In 1959 a man disappeared while vacationing on a boat sailing down the Danube near Vienna. There was a suspect, but no body had been found. Austrian investigators asked Professor Wilhelm Klaus, a palynologist at the University of Vienna, to examine pollen in mud from the suspect's shoes. In the mud Klaus found recent spruce, willow, and alder pollen, and, most interestingly, fossil hickory pollen from 20-million-year-old Miocene sedimentary rock. He knew that this combination could only be found in a small area about 12 miles north of Vienna. When police confronted the suspect with this information he confessed and led them to the burial site.

The identification of pollen and spores has numerous applications in forensic studies. Investigators have traced travel routes of drug shipments by examining accumulated pollen and spores. They find it in airplanes, trucks, and cars; air filters are a particularly good place to search for evidence. Pollen associated with bodies can help locate original crime scenes. It can accumulate on the clothing or skin or in the stomach and intestines. Many studies are now in progress to verify the country or location of origin for such products as honey, dried fruit, and coffee. Antique furniture and art has been examined to see if the pollen is consistent with the story about time and place of manufacture. Undoubtedly future use of these little particles will increase and researchers will find many new applications. More information on forensic palynology can be found at www.crimeandclues.com/index.php/physical-evidence/trace-evidence

5 Artificial and Commercial Earth Materials

We use rocks and minerals industrially and commercially in a wide variety of ways. We also manufacture and distribute artificial minerals and mineral products such as glass and abrasives. In these commercial products, minerals and rocks are often mixed with other materials, or new materials are created that are highly distinctive and therefore valuable as evidence. This is particularly true in urban areas where human products are concentrated. Indeed, these products are common alongside natural particles in the soils of our cities.

Examiners usually study these materials using the same methods they use for natural materials. Often, for forensic purposes, the fact that a substance is an artificial or commercial product is valuable in and of itself. For example, the presence of a recognized insulation material used in safes on a person who is not in the business of building or repairing safes raises serious questions as to the source of that material. Normally encased in steel, it could only be liberated by blowing, breaking, cutting, ripping, or drilling into a safe.

The identification of artificial mineral products served investigators well in a recent case in Great Britain. The case began with the murder of a woman. Police believed that a driver for a company that operated a fleet of trucks was responsible, but they did not know which driver. On the victim's clothes were many tiny specks of the "flint" material used in cigarette lighters. Examination of the company's trucks turned up one that had identical specks on the passenger seat. The driver of that truck was a chain-smoker and used a lighter. The lanthanide cerium, a metallic material from the lighter flint, apparently fell on the passenger seat and stuck to the victim when she was in the truck. The driver of the truck was arrested and convicted.

Manufacturers guard the composition of many manufactured products as trade secrets. However, most forensic laboratories know the composition of such materials. Even when scientists cannot trace a material back to a specific product, they can apply the same methods of comparison they use with natural earth materials if a sample of the material is available for control. Generally, laboratories maintain large collections of manufactured mineral products for use as control samples.

Glass

Manufactured glass is extremely common evidence in many types of crime. Broken automobile headlights, windows broken when a suspect enters a building, broken bottles used as weapons, glass beads used in projection screens and highway reflecting strips, rock and glass wool insulation material—the list seems endless. In many cases, two or more different kinds of glass have been found on suspects that compare with glass associated with an illegal entry. Multiple comparisons make even stronger evidence than single comparisons.

Most glass is composed primarily of calcium, sodium, and silica, with minor amounts of other elements. It may be natural, as in the case of volcanic glass (obsidian) or, more commonly, man-made. Glass is manufactured by melting quartz sand and other minerals and cooling the liquid. Depending on the type of glass, the product is commonly annealed by moving the glass through a tunnel that is hotter at one end and allowing the glass to slowly cool as it moves toward the cooler end. Annealing removes stresses and strains within the glass, changes the refractive index, and makes the product more uniform.

Humans have been making glass since about 4000 B.C. Modern manufacturers produce myriad types of glass. Currently there are over 100,000 glass formulations and more than 1,000 glass compositions in production. The most common glass, soda-lime glass, is used in window glass, bottle glass, light bulbs, and pressed ware. To make sealed beam headlights and heat-resistant glass, manufacturers add the mineral boron. Lead-alkali glass is used for crystal glassware, neon-sign tubing, and video tubes. Manufacturers produce tempered glass for automobile windows by rapidly heating and cooling the glass surface, thus introducing stresses. Tempered glass shatters into small squares on impact. Laminated glass, with a layer of plastic between two layers of ordinary glass, is used in many American automobile windshields. Colored glass results from the introduction of various elements into the glass mix.

In a recent case in the Rocky Mountain West, glass helped tie a suspect to a crime. A group of teenagers gathered to drink in a remote area outside a city. One of the young men tried to rape one of the women. A male friend rescued her and drove her to town. The bad guy followed and, in a rage, severely beat the rescuer with an aluminum baseball bat, smashing out the windows of his car for good measure. Very small slivers of the window glass stuck to the bat. Removed and compared with samples of the rescuer's window glass, the glass helped associate the suspect's bat with the vehicle and the beating.

In some cases, investigators can fit pieces of broken glass back together like a jigsaw puzzle, producing truly individual evidence. In most cases, the study of glass for forensic purposes depends on variations in the properties of a material and the examiner's ability to accurately distinguish small differences in those properties. The four properties or methods most useful in the forensic examination of glass are density, refractive index, chemical composition, and dispersion.

To measure density, examiners take a small glass chip and immerse it in a liquid mixture of bromoform and bromobenzene. They vary the amount of these two chemicals, adjusting the density of the liquid, until the glass chip remains suspended. Glass chips of approximately the same size and shape are then added to the liquid for comparison. Those particles that remain suspended have the same density as the initial control particle. This flotation method can adequately distinguish glass particles that differ in density by just a milligram per milliliter.

The refractive index is commonly measured using one of two different methods. Some laboratories use the Emmons double variation method. In this method, examiners place glass fragments in a liquid—usually a silicon oil with a refractive index that varies with changes in temperature on an electrically heated microscope stage. Refractive index is also a function of the wavelength of light, so the microscope has an attachment that varies the wavelength used to view the specimen. When the examiner can no longer see the glass fragment, that means its refractive index is the same as that of the oil. Measurements are made at three different temperatures and the data plotted on a Hartmann net. The plot makes it possible to determine the refractive index of the glass fragment at any wavelength of light. The second way to measure the refraction index, used by many labs today, is the GRIM (Glass Refractive Index Method) II method. GRIM II uses a video camera and imaging technology to detect the match point. A hot stage varies the temperature; filters fix the light wavelength; and computer software calculates the refractive index with values reliable to 0.00007. This method reduces operator error, but dispersion data are difficult to obtain. Dispersion is the difference in refractive index when measured in blue and red light. Today, with modern glass manufacturing methods producing more uniform products, dispersion is losing value in the differentiation of glasses. However, as long as old glass is still around, dispersion measurements will have a place in forensic examination.

Chemical analyses of glass can provide useful information in some cases. Variation among most of the major elements is insufficient for their values to be of much use. However, the presence and amount of trace elements may provide significant discriminating information. Examiners use many kinds of analyses, foremost among them inductively coupled plasma emission spectrometry.

In a classic case, an individual insulated his home with rock wool he bought at several different sales. That meant that the materials came from several different manufacturers and were produced at a variety of times. An intruder entered the home and crawled through the attic, picking up particles of the insulation on his clothes. Later, these particles compared with the unique combination of insulation materials in the attic, providing strong evidence that the suspect had been in the attic.

Safe Insulation

Generally, we use two types of safes in the United States: fire-resistant safes, which have 2 or more inches of insulation between their walls, and burglar-resistant safes composed entirely of metal (often encased in concrete or placed inside a fire-resistant vault or larger fire-resistant safe). Because the insulation in fire-resistant safes is often porous and relatively soft, it commonly turns up on the clothing and tools of the safe cracker, or in his toolbox or vehicle.

Manufacturers have used numerous materials for insulating safes, including dolomite and cement; cinders; corrugated asbestos; gypsum with cork, asbestos, or wood chips; gypsum and/or calcium carbonate with diatomaceous earth; and even ordinary mud or cement.

Today, the insulation in safes made by some of the leading manufacturers contains Portland cement, vermiculite mica, and diatomaceous earth. If Portland cement, vermiculite mica, and diatomaceous earth are identified together, examiners can reasonably conclude that the material is safe insulation, since few other products have just those three ingredients.

Many safes made before 1936 contain natural cement as insulation. Natural cement is a product resulting from the calcining of certain argillaceous limestones. The argillaceous material, usually shale, is easily observed under a low-power microscope and varies in color, particle size, and abundance. Because some natural cement was produced by burning

This safe was opened by ripping. Chunks of insulation and powdered insulation material are on the floor, safe, furniture, and almost certainly on the clothes of anyone who was there while the safe was being ripped open.

limestone and coal or coke in a kiln, insulations made from this product will contain coal particles and cinders. Since there is no record of this insulation being used without sand or gravel in anything but safes, even a small lump of such natural cement tells the forensic microscopist that it is insulation from an older safe.

Some modern lightweight safes use a very porous mixture of Portland cement and vermiculite. One large company, under several brand names, produces safes using an insulation composed of gypsum and sawdust; another uses gypsum alone.

From experience and by maintaining files of various insulations, the forensic microscopist can identify many safes by examining insulation. Because of the variations in insulation, comparisons of safe insulation from crime scenes and suspects' clothing are always valuable.

Diatomaceous earth, sometimes known as Kieselguhr or infusorial earth, can yield evidence of value because of the diversity of diatoms in most deposits. Diatoms are unicellular organisms that produce silica shells. There are over 100,000 species of diatoms, both modern and fossil, marine and freshwater. In one case, what appeared to be a bad case of dandruff in a suspect's hair turned out, under a powerful microscope, to be thirteen distinct species of diatoms. Examination of the insulation of a recently breached safe revealed that the diatomaceous earth it was made of contained the same thirteen species of diatoms. This evidence led to the suspect's arrest and subsequent conviction. In 1980, however, most manufacturers discontinued the use of diatomaceous earth as an insulation material in safes.

One night in 1997 three officers of the police department in Riverside, California, encountered a man they judged to be drunk. For reasons that are not clear, they beat him and threw him into a man-made lake in Riverside. The victim complained to the police and an internal investigation was begun to determine if there was any physical evidence that corroborated or did not corroborate the victim's story. The Riverside police department contacted Marianne Stam, forensic geologist with the Riverside Crime Laboratory of the California Department of Justice. Stam has had a long record of success in her work because of her ability to think out problems and come up with creative approaches to her casework. The officers' clothes had been cleaned and thus offered no evidence. Stam examined the victim's clothes and saw water staining on them. She set out to determine if the water staining came from the specific lake in question. Soil samples from around the lake and water samples from the lake contained diatoms. She collected samples from the victim's clothes using sticky tape, then placed the tape on microscope slides and covered them with ultra-pure water. Under the microscope these samples showed abundant

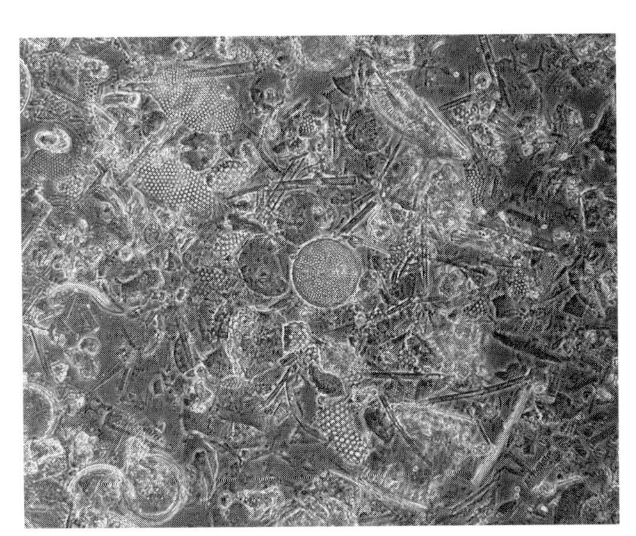

Diatomaceous earth magnified approximately two thousand times
—COURTESY OF MIKE CLAYTON, UNIVERSITY OF WISCONSIN PLANT IMAGE COLLECTION

diatoms. Stam contacted an expert on diatoms who confirmed that the diatom population in the water samples from the lake and the samples from the victim's clothes were identical. The victim's clothes had been in contact with the lake water and thus it appeared the victim had been in the water. In addition, the soil samples and the victim's clothes contained parts of the Mexican bald cypress tree. These trees were imported in the early 1900s specifically to be planted around the man-made lake, with one tree also planted in a botanical garden 10 miles away. Therefore, the presence of the plant material on the clothing, in addition to the apparent water stains and diatoms, provided evidence that supported the victim's statements to police investigators.

Stam presented her findings to the internal investigators. Faced with this evidence as well as other investigative information, the officers confessed and were fired.

Asbestos, the "miracle mineral," has excellent fire-resistant properties and was commonly used as safe insulation. Asbestos is the fibrous variety of several of the amphibole minerals or more commonly the fibrous serpentine mineral chrysotile. The various types are easily distinguishable. Insulation materials differ according to manufacturer, era of manufacture, and source of material. However, with the discovery of the health hazards of these fibrous minerals, use of asbestos in modern safes has become limited.

In most cases it is possible to use standard methods of comparison for samples from broken safes. A case from a small town in southern Maryland provides a typical example. Two safes were broken into, one in a movie theater, the other in a nearby restaurant. Police apprehended two suspects shortly after the restaurant crime. One admitted to the crimes and said that the other was completely innocent. However, a cement-type safe insulation in the trouser cuffs of the supposedly innocent suspect compared with particles of insulation from the restaurant's broken safe, implicating him in the crime. Particles of a vermiculite mica safe insulation found in the suspect's automobile also compared with the insulation in the broken safe from the theater, associating him with that crime as well.

Building Materials

Bricks, concrete blocks, plaster, cement, ceramic materials, roofing granules, and other building materials can offer important evidential value. Fragments of these materials commonly adhere to clothing or tools during forcible entries. They also become embedded in the fenders and frames of automobiles during impact. Examiners have even studied minerals adhering to fired bullets, in many cases, to determine if a bullet ricocheted off building material. Because construction materials are commonly made at a specific time from mineral materials combined for a specific purpose, they can be highly distinctive. Regional differences in the minerals and rock in building materials can be very useful. For example, as seen in earlier chapters, manufacturers in different parts of the United States use different raw materials to produce concrete blocks. An examiner might find sea shells in a block manufactured in the South, artificially produced cinders from furnaces in one from the East, and natural volcanic glass in concrete made in the Northwest.

Concrete can be important evidence. No two batches of concrete are identical. Texture, mineralogy, and particle kind and size vary from batch to batch due to the amount of water used to mix the concrete, the amount and kind of aggregate added, and a host of other factors. It is almost impossible to mix concrete the same way with the same materials twice. Examiners can learn many things from concrete. They can testify with certainty that two concrete samples with slight differences in properties did not come from the same batch. They can testify that two samples of concrete with identical properties have a very high probability of coming from the same batch. Most important, if two samples of concrete have some but not all important properties in common, examiners know that the samples may not necessarily have come from anywhere near one another.

During World War I, forensic geology and diplomacy came together in a most interesting case. Britain was involved in a dispute with the Netherlands over whether the Dutch were allowing the Germans to transport military supplies and concrete aggregate on Dutch canals in violation of their neutrality. Between September and November 1917, 300,000 to 400,000 tons of sand and gravel moved through the Netherlands by canal to Belgium. The claim was made that the material was for nonmilitary purposes. Captain W. B. R. King, geologist to engineer-in-chief G. H. Q. France, took on the task of determining the source of the concrete aggregate in German pillboxes captured by Allied forces. He tested thirty-nine samples of concrete collected at Vimy Ridge and the Ypres Salient. Examiners found a notable absence of Belgian rocks; thirty-two of the samples contained material from sources outside Belgium. Some samples included such uniquely German rocks as Niedermendig lava and Niedermendig tephrite. Triassic sandstone in thirteen samples occurs as pebbles in the Rhine valley, but not in Belgium. Informally, the British sent this information to the Dutch government. In recognition of the difficult, fragile position the Netherlands were in at the time, the British did not press the matter further, but the German lie had been proven.

More recently, brick and plaster provided evidence in a case of illegal entry of a national organization's office in a town in eastern Idaho. The safe was blown using nitroglycerine and the contents, amounting to several hundred dollars, was removed. Entry was made by removing brick and plaster from part of a wall. In a nearby hotel room, particles of brick and plaster that compared with those from the crime scene suggested that the room's former occupant should be considered a suspect. Police apprehended the suspect in another city while he was casing another office of the same organization.

Manufacturers produce plaster by heating the mineral gypsum to slightly over 100 degrees Celsius (212 degrees Fahrenheit), driving off some of the water contained in the mineral. When water is added back into the plaster, new crystals of gypsum form, setting or turning the material from a powder into a rocklike material used in wallboard, safe insulation, and other building materials. Identification of the rock and mineral particles in plaster and cement is an important aspect of studying these materials. When they are collected for forensic examination, it is important that they be placed in an airtight container and not heated again, thus preventing the conversion of all or part of the gypsum back into its plaster or low-water form. Positive comparison would be impossible in such a case. Even accidental heating during preparation of a thin section can destroy evidence.

After an arson, investigators attempt to collect from the burned area a sample of the unburned fuel used to accelerate the fire. If the suspect had access to the same type of fuel, the sample can serve as evidence. In Japan an arsonist tried to hide the fuel he used by making a small hole in the outer wall of a stucco house with a Phillips screwdriver, pouring a small amount of fuel into the hole, and lighting a match. However, when investigators found the screwdriver in the suspect's shop and tested the residue on it, fragments of rice plants, gypsum, and other stucco minerals in the material compared with the stucco covering the house. The information contributed to the arsonist's conviction.

In the Netherlands, government regulation has resulted in a valuable forensic resource. In 1999, to encourage recycling and reuse of building materials, the Dutch government issued the Building Materials Decree. The decree sets rules for the use of building materials, both new, such as sand, gravel, and limestone, and recycled, for example, residue from industrial processes and demolition waste. Because differentiating between new and recycled building materials can be difficult, the Netherlands Forensic Institute, which helps enforce the decree, has conducted extensive studies of

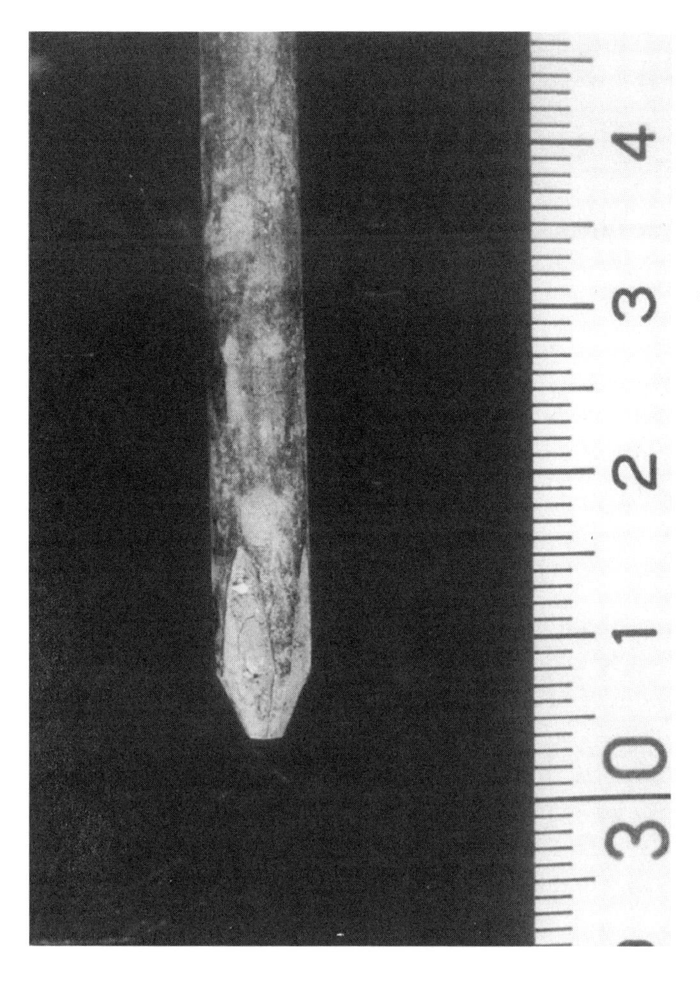

Phillips screwdriver used by arsonist to drill hole in stucco wall. He put fuel in the hole and started a fire. Building material on the screwdriver was used as evidence in court. Scale in centimeters. —COURTESY OF YOSHITERU MARUMO, NATIONAL RESEARCH INSTITUTE OF POLICE SCIENCE, JAPAN

building material properties to assist in identification. The resulting base of information has helped identify materials associated with other crimes as well.

Cleaning and Face Powders

Commercial powders used for cleaning and in cosmetics commonly have a mineral base or mineral filler. The specific minerals used differ from one product to another, and manufacturers tend to change the size and composition of minerals through time. Thus, these materials can be studied for comparison or lack of comparison. In the case of a man who assaulted a young woman by kicking her in the face, the identification of powder on the toe of his shoe and the comparison of that powder with a sample of her face powder proved convincing. X-ray diffraction analysis identified a titanium oxide base in the form of the mineral anatase in the face powder. It is interesting to note that a chemical analysis would have detected the titanium, but that better evidence was produced by identifying the specific titanium mineral.

The soft mineral talc forms the base of most talcum or dusting powders. These powders vary in particle size from product to product; some are ground more finely than others. They also differ in the associated minerals that are found with the talc. Different manufacturers commonly obtain their talc from different mines, thus providing the possibility of different minerals associated with the talc. In addition, manufacturers' formulas change with time, producing considerable variations in the products. For example, since 1972, cornstarch has seen increased use.

Abrasives

Abrasive materials may be natural—such as emery, which is a mixture of the minerals corundum and magnetite, hematite, or iron spinel—or artificial, such as carborundum and alundum. Natural diamond, garnet, and various forms of quartz are also found in some abrasives. Natural materials tend to be more diverse than artificial materials. Both kinds have forensic applications. Abrasives have served as evidence in cases involving malicious damage and sabotage, where the material was introduced into machinery, causing damage to the working parts. Grinding wheels used to remove serial numbers or to open safes sometimes leave particles of grit, offering the possibility of associating a suspect with the crime.

Commercial Sands

A wide variety of sands are used for commercial purposes. These vary from ordinary quartz sand sold as it comes from the sand pit for such purposes as concrete aggregate, to a variety of special sands that may be treated to remove certain sizes of particles or minerals, or blended with other sources to produce a product for the manufacture of glass, oil well treatment, sandboxes, molding forms used in foundries, and the like.

Sand helped associate a suspect with a crime in a case of the breaking and entering of a foundry in Toronto, Canada. In the case, the suspect's shoes contained abundant grains of the mineral olivine. This mineral does not exist naturally in Toronto-area soils, but three foundries in the area imported it as a special molding sand. Some was found spilled outside the window used for the entry. Though the suspect denied ever having been at any of the foundries, the olivine sand from his shoes compared with the sand collected from outside the window. Combined with the known distribution of potentially similar material, the mineralogical comparison made it highly probable that the two samples came from the same spot and that the suspect had been at the scene.

Potting soil is a common commercial product with diverse formulations. These can include minerals such as volcanic ash and expanded vermiculite, plant material, and even plastics. In a case of rape that occurred in upper Michigan, the quality and significance of the geologic evidence was overwhelming. In the struggle that took place, three flowerpots were overturned on the victim's living room floor. The suspect had potting soil on his shoes that compared in all details with soil from one of the three pots, but not the other two. This sort of differentiation is possible because potting soil is not a product that generally undergoes extensive quality control. Differences commonly exist from bag to bag, making the match between the sample from the suspect's shoe and the overturned pot all the more valuable. In this case, additional evidence was significant. Soil samples from both the suspect's shoe and the living room floor also contained small clippings of blue thread identical in size, color, and other characteristics. A reasonable person would conclude that the suspect's shoes had been in that living room after the flowerpot had been overturned, placing the suspect at the crime scene.

6
Evidence Collection

All readers of detective stories or watchers of police or forensic programs on television know that the first thing that happens at a crime scene is a call to the forensic lab people. Soon a van arrives, filled with masses of equipment and several grim-faced evidence collectors. In reality most crime scene evidence is collected by the investigators or by an individual trained and assigned to collect evidence. That person may be employed by the law enforcement agency or forensic laboratory that will process the evidence. In addition, around the country, the FBI maintains groups of highly skilled evidence collectors who can rush to high-profile or massive crime scenes such as the Oklahoma City bombing.

In most forensic applications of soil studies, there are two basic types of samples. The first are samples directly associated with the crime or incident—the questioned samples. As we have seen, these come in many forms: lumps of soil on a highway at the scene of an accident, soil on shoes or clothing, safe insulation, dust in hair, microscopic marine or freshwater creatures in the lungs of a drowning victim, rocks or glass used as weapons, abrasives in machinery, and so on. The second type of sample, which the investigator or forensic geologist can select, is the known control material. Examiners compare these with questioned samples. Control samples can include soil removed from the frame and fenders of a suspect vehicle, soil from a crime scene, insulation samples from the victim's safe, water samples containing microscopic animals and plants from suspected areas of drowning, rocks or glass from places a weapon may have come from, sources of abrasives, and so on. Known samples can also consist of material from museums or other collections, maintained for the purposes of comparison.

Questioned or Associated Samples

Questioned, or associated, samples are normally acquired by accident, with no attempt to provide a good representative sample. For example, a rapist rarely chooses the best sample of soil for his trouser cuffs, a sample that is the most representative of the soil at the scene of the crime. What lands in the trouser cuffs may lack some of the larger particles present at the scene, for example. Such a sample can never be expected to be exactly the same as a known control sample, which includes all the available sizes. In such cases, the examiner can only study the particles in the control sample that are the same size as those in the questioned sample. In a case of forcible breaking and entering, earth materials collected from entry tools or clothing—roofing granules, for example, or masonry, brick, or rock—may not be truly representative samples. During such an incident, a person may collect just those grains that are loose or most easily broken. It is unlikely that these would be the same as a bulk sample of the original materials. In such cases, the examiner must use his or her professional judgment in choosing the best methods of analysis for establishing comparison or lack of comparison. The competent professional can make such judgments; the less-knowledgeable examiner may blindly attempt to process the bulk samples without judgment.

A second type of sampling for most associated or questioned material involves the forensic geologist when he or she removes earth material from the shoe, clothing, vehicle, shipping box, or the like. Where a lump of soil is involved, it should be collected and preserved intact. Preservation of the original sample is especially important when layers of material are involved, as with a lump of soil from underneath the fender of a vehicle. Such samples permit the study of stratigraphy, that is, the particles of individual layers from oldest to youngest. From such samples one can also see how the particles that make up the lump fit together. After searching for intact lumps, examiners collect samples from clothes by shaking them over a clean sheet of paper and carefully gathering the debris. Sometimes they use a spatula to dislodge material or a vacuum cleaner with a clean collecting bag. However, vacuum cleaners can break lumps of soil and change the physical appearance of particles. In 1951, Max Frei-Sulzer of the Zurich police gained notoriety when he began using Scotch adhesive tape to collect and preserve small samples. While this method has many advantages, the tape may interfere with some analyses and it is sometimes difficult to remove particles from the tape.

The forensic geologist must always be aware of the context and source of the samples examined. In a recent case of armed robbery, the suspect wore a ski mask. The crime lab submitted a sample of "soil" from the mask for comparison with soil samples from the crime scene. The forensic examiner working on the case found that samples from the mask and from the scene matched in all aspects except for an abundance of glass beads in the sample from the mask. Had the sample somehow been contaminated by glass beads? Hours were spent examining the sample's chain of custody. The mask could have been stored near projection screens or highway markers, both sources of glass beads. But no possible sources of contamination were found. What the examiner did not know was that the ski mask had a strip of reflecting tape across the forehead embedded with glass beads and that these had been included in the soil removed from the face mask by the crime lab. Had the examiner collected the sample from the face mask himself and thus known the context of the sample, the problem would never have arisen. It has been argued by some that justice is better served if the forensic examiner has no knowledge of the available information about the crime. Using this idea the examiner would only be given the samples to be analyzed. However, there are many cases like the one above that demonstrate that such an approach can lead to very bad mistakes.

Known or Control Samples

Known or control geologic samples are of two types, those collected from a crime scene or alibi location and those that exist in museums or collections as part of the scientist's professional resources. Samples from crime scenes or alibi locations may be collected in at least two ways. Usually, investigators or evidence collectors collect samples as part of the investigation and submit them with other items of physical evidence to the forensic laboratory. In this case, the responsibility for proper sampling lies with the collector, who uses procedures normally determined and published in instruction manuals by the laboratory. Instruction manuals are usually available to potential clients.

In the case of materials from vehicles, separate samples should be taken from under all four fenders, with any intact lumps of soil preserved. In accident cases, collectors should gather oil or grease, including any minerals, rocks, or related material mixed with it, from several places under a vehicle if these are to be compared with similar material from the scene. Samples should be carefully labeled with the location where the sample was taken. When the crime scene includes a vertical cut into the earth—a quarry, for example, or a grave—samples should be taken from each bed, layer, or horizon that exhibits a visibly different color, texture, or mineralogy. When glass, building materials, or synthetic materials are involved, the sample should be pure, sufficient for analysis, and representative. Two or more different pieces of broken glass present at a scene should be sampled separately.

Collecting evidence at the scene of a hit-and-run accident

Soil on the tailpipe of a suspect's vehicle

In many cases, investigators collect control samples simply by picking up soil in the area of the crime or alibi location. We have already seen that the make-up of soil and related material varies markedly within even very short distances. Considering how slim the chances are that two samples from inches away will share the same properties, it is just as unlikely to find control samples with the same properties as a questioned sample. It is better practice for the evidence collector to examine a questioned sample first, for color and particle size at least, and then to search for samples with a similar appearance. The failure to produce significant evidence in the laboratory is often the result of sloppy sampling. One hundred samples that do not compare contribute no significant evidence and may only lead to frustration and lack of trust in the examiner. The search should lead to samples that compare, if such material exists.

Here is an analogy. You want to know if a person has been to a barbershop and had a haircut today. You know that the barber has not swept the floor of his shop since he opened. The person in question has red hair. If you randomly collect 100 hairs from the floor of the shop, you might or might not pick up a red hair. If you did pick up a red one, it might or might not have the same properties as the hair of the person in question. By selecting only red hairs from the floor, you increase the possibility of finding a hair that compares with that person's. If you find a hair that compares, you have answered the question. The person got a haircut today. If you examine all the red hairs and find none that compare, you have shown that the person did not get a haircut. Or let's say you're looking for a pair of socks. You have one sock with blue polka dots. You want to find a match in a drawer filled with all kinds of socks. If you randomly take ten socks from the drawer, you might or might not make a pair, even if a matching sock exists in the drawer. But removing only socks with blue polka dots increases the chances of finding a matching pair, if such a pair exists.

Control samples of minerals or fossils from museums or other collections—a normal part of the forensic geologist's professional resources—can help examiners identify questioned materials and locate places where similar material could be obtained. Geologic maps of the surface of the earth showing the distribution of kinds of rocks are invaluable in determining possible sources of materials. In a case of assault and vandalism in which people were injured by hurled rocks, the rocks recovered at the scene were angular blocks of an unusual igneous rock that exists in a relatively small area in the southern United States. Study of geologic maps indicated that the nearest outcrop of this rock type was 20 miles from the crime scene. The rocks must have been carried to the scene. Further investigation

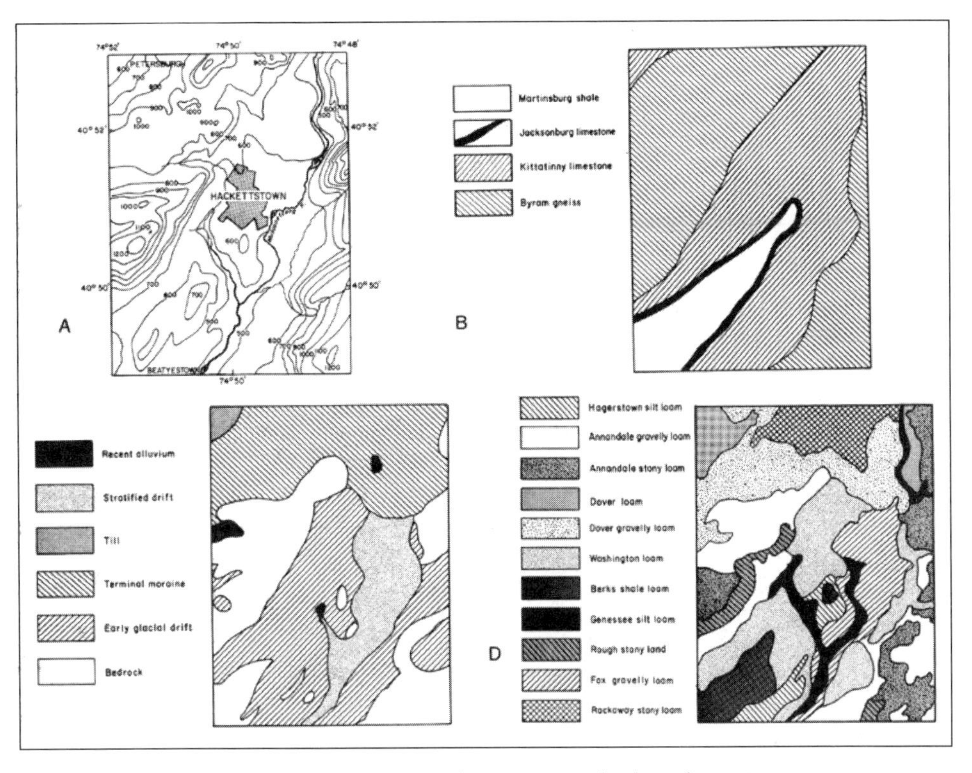

Geologic Maps. A, Topographic Map; B, Bedrock Geologic Map; C, Surficial Geologic Map; D, Soil Map

in the area of the rock outcrop produced a suspect who drove a pickup truck and carried similar rocks to add weight to the truck.

Topographic maps showing land elevations can be helpful. In one case, an informer told agents of an alcoholic beverage commission about an illegal distillery between two towns near a railroad. The informer said the site had a well whose water level was 20 feet below ground. The last point may seem to be an unusual piece of information, but would be common knowledge for a well driller. A topographic map of the area revealed many swamps, as well as ridges of sand and gravel. Investigators assumed that the water level in the ground, and thus in the well, would not be much higher than the elevation of the water table in the swamp. Only one place was 20 feet or higher above surrounding water levels, and that one possible location was occupied by a church. Further investigation revealed the illegal still in the church basement, and led to successful prosecution and destruction of the still.

Collection of Soil Samples

Because of the layering effect in natural soils, samples should be collected by horizon. Most questioned samples come from surface layers, however, so only top layers are needed for comparison. Soil samples should be placed in clean plastic cartons, plastic vials, or other leak-proof containers. Wet or moist samples should be air-dried before they are placed in containers. Otherwise, biological activity will continue; certain organic components may break down and other new ones may form. Conversely, samples collected for conductivity measurements or that contain volatile substances should not be dried but should be sealed and refrigerated until analyzed. The amount of sample required for analysis depends on the type of study. Most analyses require approximately one cupful of soil. However, for mechanical analysis of particle sizes in soil, examiners may need a larger sample. If considerable gravel or other coarse material is present, the size of the sample should be increased accordingly. However, if large quantities of a sample are not available, a few grains of soil are better than none.

Drying alters some soil properties but not others. The main properties of a soil that may be subject to change upon drying are as follows:

- If salts are present in a sample, the salts will be concentrated and may crystallize on the surface of the sample.
- Some minerals may oxidize or be subject to other alteration and, in so doing, may change color. This is especially true of black sulfur-bearing muds from swamps or marshes.
- The nitrate content of dried samples tends to increase.
- Populations and activities of microbes may alter greatly.
- Soil samples tend to lighten in color upon drying.

In the case of soil samples collected from certain wet areas such as marshes or bogs, it may be desirable to seal them in plastic or glass containers and refrigerate them. This will help prevent changes due to microbial activity or oxidation.

Water samples should also be collected in plastic or glass containers, sealed, and refrigerated. The low temperature will retard biological growth and prevent important changes. For pollen studies on geologic or soil samples, the examiner must take care to prevent contamination from other pollen in the air throughout the year. Specimens and samples should be sealed in plastic wrapping. In analyses of trace metals, there is always the question of metal contributions to the sample from digging implements, metal containers, and sieves used in collecting and storage. It is preferable to use plastic or glass storage boxes and stainless steel implements and sieves.

Two principles are vital in evidence collection. First, samples must be collected by legal means if they are to be admitted as evidence. That usually means with proper permission, warrants, or incident to an arrest. Geologists love to collect rock samples and have even been known to jump fences for an unusual rock. When dealing with evidence in a court of law, however, this won't fly, no matter how noble the cause or temptingly easy the procurement. The second principle has to do with maintaining the chain of custody. This means keeping a written record each time material is transferred, and ensuring that each person who has custody of a sample is responsible for its control until it is transferred again.

When scientists see that important information might be obtained from additional samples or information, they may arrange to obtain these samples or information. For example, in a case of forcible rape in an area of glacial outwash, the suspect was found to have a large accumulation of sand within his trouser cuffs. The victim gave the investigator the location of the scene of the crime, and the investigator went to the site and collected soil samples. The investigator found that the samples of sand from the suspect's cuffs compared with samples from the scene. This comparison was made on the basis of ten unusual rock types existing as particles in both samples. In addition, no rock types were found in one sample that did not exist in the other. One of the unusual rock types was anthracite coal. It was known from geologic experience that anthracite coal did not exist naturally in the area and also did not exist north of the area and thus could not be a natural particle in glacial outwash. Despite this, the coal was common in the samples, making up over 5 percent of the particles. The investigator established that the scene had been the site of a laundry sixty years earlier. The coal in the soil was apparently the remains of a former coal pile. This additional information further increased the already high probability that the soil in the suspect's trouser cuffs had been picked up at the scene of the crime.

7 Examination Methods

Forensic geologists rely on various instruments, methods, and procedures to study minerals, rocks, soils, and related materials. They use them to collect data and to judge whether questioned and known samples compare. Some of these methods and instruments help provide better evidence than others. Because of the extreme diversity of soils and related material, the examiner will decide which method or methods will best assist in reaching a conclusion in a particular case. Because soils often have added particles, such as fibers, hair, or paint chips, these can be collected for examination by experts in those forensic science fields. Such particles greatly strengthen the determination that two samples have the same properties and may come from a common source, as does the presence of rare or unusual minerals.

Microscopic examination is extremely important. It is the best way to find particles of foreign matter or unusual minerals in samples. Although we may not be able to provide conventional statistics regarding the probability of a match between samples, the forensic geologist can express a reliable opinion based on experience and education. That opinion is grounded in knowledge of the rarity of particular minerals, rocks, and particles, and their related properties. It is strengthened when an examiner has never before seen a sample like the one before him or her.

Color

Color is one of the most important identifying characteristics of minerals and soils. In the 1970s investigators at the Home Office Forensic Science Service studied the use of color as an examination method. They helped establish the study of color as an important first step of examination. More recently, Y. Marumo and R. Sugita of the National Research Institute of Police Science in Tokyo have added to knowledge of this important tool.

Together, minerals form a mosaic of grays, yellows, browns, reds, blacks, greens, brilliant purples, and more, representing virtually all the colors of the visible light spectrum. With most geologic materials and soils, the native minerals contribute directly to the color. This is particularly true with stream deposits, windblown silts, and other formations that have been in place for a comparatively short period of time. If you look at sands along a river channel, the color of each sand grain, generally, is individual. However, a long period of weathering results in a degree of leaching, accumulation, and/or movement of substances within the soil. Soil particles become stained, coated, and impregnated with mineral and organic substances, giving the soil an appearance different from earlier ones. Larger mineral grains in particular become coated. Most often, coatings on soil particles include iron, aluminum, organic matter, and clay. The coloring of coatings alone can shed light on the history of a sample.

Redness in soil not only depends on the amount of iron present, but also on the iron's state of oxidation. A more highly oxidized condition results in a deeper red. Iron in coatings on particles most often occurs in the form of hematite, limonite, goethite, lepidocrocite, and other iron-rich mineral forms. Black mineral colors in the soil are generally related to manganese or various combinations of iron and manganese. Green colors are usually caused by concentrations of minerals rather than minerals in coatings. For example, some copper minerals, chlorite, and glauconite are usually green. Deep blue to purple coloration in soil is generally due to the iron phosphate called vivianite. Apart from the colors caused by minerals, some colors in soil result from organic matter. The organic litter on the soil surface is generally black. Humus percolates through the horizons of mineral in soil, producing various dark colors. In some instances, iron and humic acids combine to make a dark reddish brown to nearly black.

To describe the color of geologic materials and soils with some uniformity, professionals use certain standards—most frequently, in the United States, those of the Munsell Color Company. These are based on three factors: hue, value, and chroma. Hue is the dominant spectral color, value is the lightness color, and chroma is the relative purity of the spectral color. As an example, a soil or rock color is recorded as 7.5YR5/2 (brown). The 7.5YR refers to the hue; the second 5, the value; and 2, the chroma. Moisture content also affects soil color. If a soil is dry, it may look yellow; if moist, it may be recorded as yellowish brown. Adding moisture to a dry soil usually adds brilliance to its appearance. It is therefore important to record not only a soil's color, but also an estimate of its wetness. Light intensity and wavelength also affect soil color. Color may vary in natural, fluorescent, and incandescent light.

Since soil is a mixture of materials of various sizes and compositions, it also contains individual minerals of different colors. If soil is fractionated into various sizes—coarse sand, medium sand, fine sand, silt, and clay—the finer-sized particles tend to exhibit more red or reddish brown colors as opposed to the grays and yellows of the coarser fractions. The matrix of coarser sand particles commonly has a speckled appearance, with gray or yellowish quartz and feldspar particles and generally black particles of such heavy minerals as ilmenite and magnetite. Sand particles in soils of recent origin, such as recent glacial or stream deposits, usually retain their original mineral appearance and mosaic of colors. But sand fractions from old landscapes often have coatings of clay, and the sand grains may be iron-stained, so the entire matrix is a more uniform color. Soil grains veneered with organic matter look dark gray. In such cases it is important to record the color of the untreated soil sample first, then to remove the organics to reveal the grains' true color and appearance.

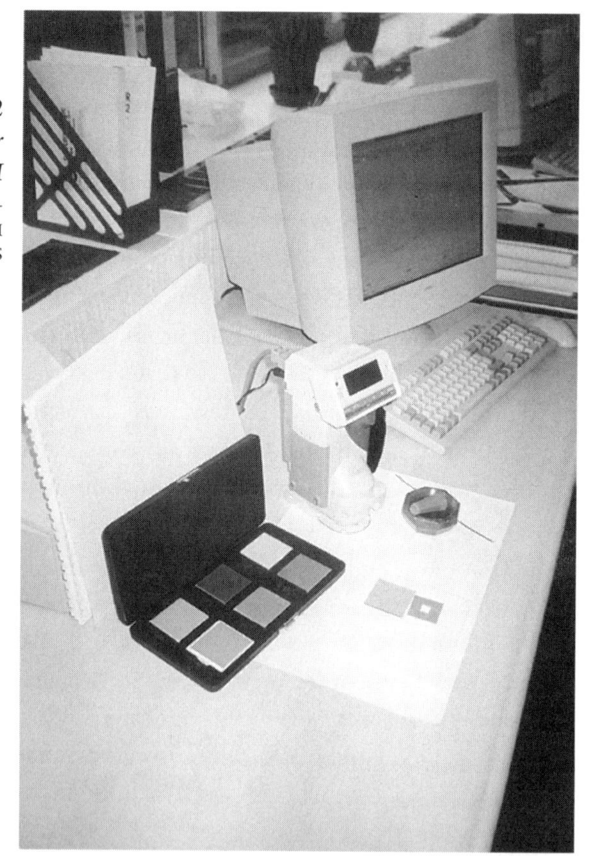

Minolta CM-2002 photospectrometer and CERAM II standards —COURTESY OF KENNETH PYE ASSOCIATES

To study soil, forensic geologists usually dry samples at approximately 100 degrees Celsius and view them in natural light, ideally from a north-facing window. The samples should have generally the same size distribution of particles. The color of individual samples sieved out according to particle size provides important additional data. Two or more samples, collected for study, can be compared directly by the observer. It is then possible, if the examiner wishes, to use a color chart with the samples to determine the Munsell color numbers for precise description of the color.

Instrumental determination of color provides more quantitative information and can now be done with precision that exceeds the human eye. Pye Associates of Great Britain achieves rapid, reliable, and highly reproducible results using the Minolta CM-2002 photospectrometer. This instrument covers the visible wavelength range from 400 to 700 nm (nanometers). Guedes and Associates in 2010 studied and reported on the use of instrumentally determined color for forensic purposes. We may see more use in crime labs of instrumentally determined color in the future.

Particle Size Distribution

Determining the distribution of particle sizes in a sample can lead to significant evidence. Examiners establish the particle size distribution in samples for a variety of reasons. Sometimes they produce samples for comparison studies that are similar. The control sample may contain some larger or smaller particles that are not present in the questioned or associated sample, in which case those particles are removed. To perform mineral or color studies, forensic geologists sometimes break samples down into subsamples in which all particles are in the same size range. And sometimes particle size distribution itself is a factor in comparison. In a court setting, a diagram of the distribution of grain sizes can have evidential value. For example, in cases of sabotage of machinery with abrasive particles, particle size distribution may help identify the material, assuming particles did not change in the machine.

The basic methods used for the separation of particle sizes are:

- passing the sample through a nest of wire sieves, with the size of the openings decreasing from top to bottom
- determining the rate of settling of the grains in a fluid, which is a measure of the size of the particles
- using instruments that measure the size of particles in a microscopic view and record the number of particles of each size

After particle sizes have been separated the data is plotted on a diagram.

Sieve and shaker with standard graphical output.
—COURTESY OF KENNETH PYE ASSOCIATES

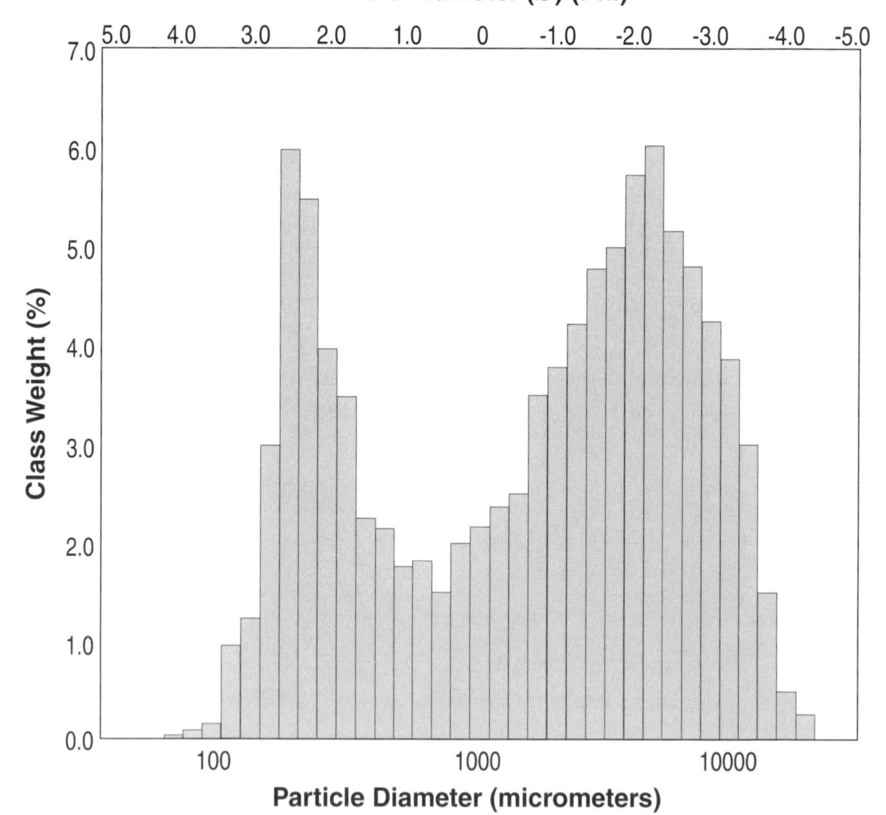

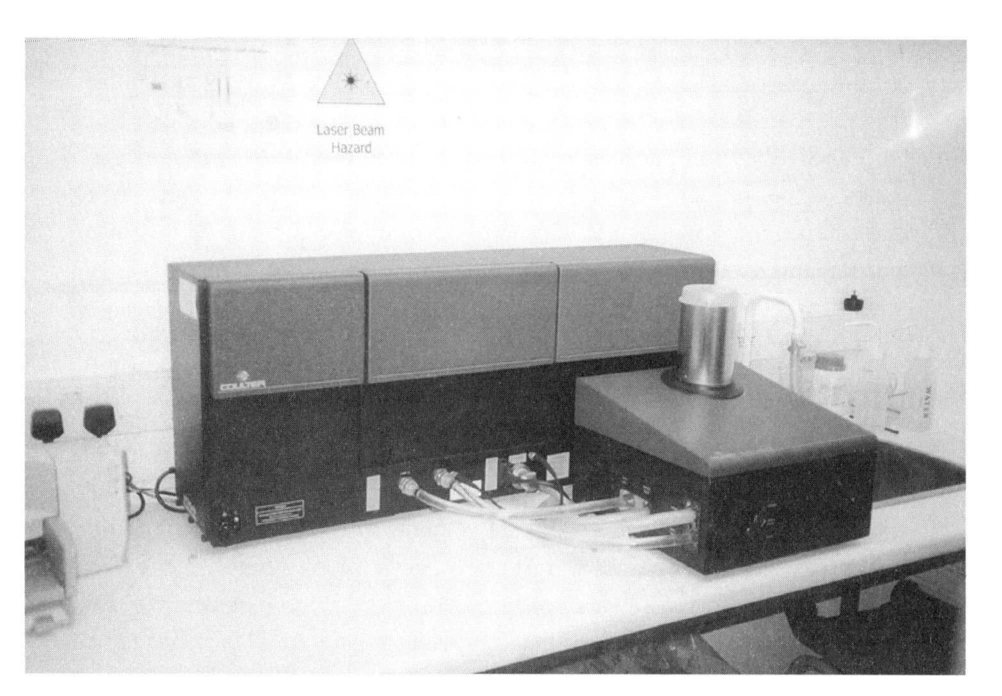

Coulter LS230 laser granulometer with standard graphical output
—COURTESY OF KENNETH PYE ASSOCIATES

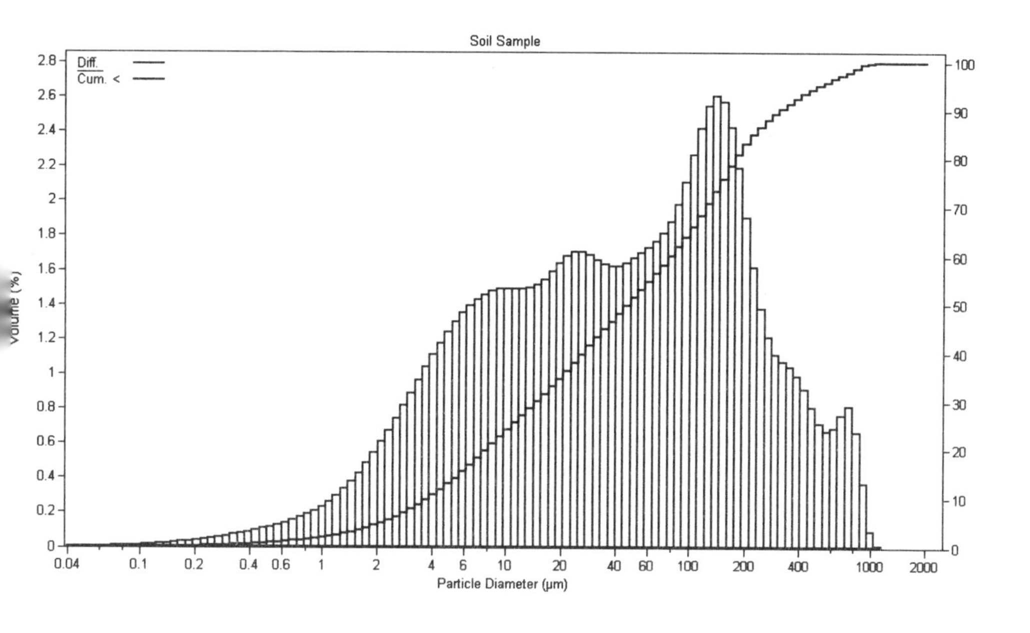

Before a forensic geologist can make a mechanical analysis to determine the size distribution of particles in a sample, it is necessary to disperse the soil. Individual soil particles tend to stick together in the form of aggregates. The examiner must remove cementing agents that hold aggregates together. Cementing agents consist of accumulated carbonates, organic matter, and iron oxide coatings. In addition, in some situations, physicochemical forces result in mutual attraction of particles.

If carbonates have cemented the particles together, it is desirable to pretreat the sample with dilute hydrochloric acid to remove the carbonates. The sample is then treated with hydrogen peroxide to remove the organic cementing agents. All samples must be treated in the same way, and it must be determined before treatment that important information will not be lost, such as dissolving carbonate cement from grains that should be treated as single grains. It is almost always desirable to determine the size distribution of soil by sieving in a liquid, usually water. Dry sieving of the entire sample is generally unsatisfactory because the small particles tend to cluster together, and clay tends to adhere to larger particles. Sometimes a dispersing agent is added to the water.

A number of methods can then be used to determine the size distribution of finer particles in a dispersed suspension. The hydrometer method determines the percentage of sand, silt, and clay in a sample. It is based on the principle of the decreasing density of the suspension as solid particles settle out. While rapid and accurate, this method is unsatisfactory if the examiner subsequently wants to examine size ranges, because there is no actual physical separation of the various-sized particles.

One of the most accurate and satisfactory procedures for fractionating soil samples is the pipette method. This consists of pretreating the sample as in the hydrometer method, dispersing the soil in water, and calculating the time various-sized particles take to settle out. Here, the underlying idea is that the rate of settling depends on the size and density of the mineral matter, with larger, denser particles settling at a more rapid rate. Although this method is generally considered the most accurate, it is not infallible. It is based on several assumptions: that all particles have the same shape and that all the soil particles have the same density, neither of which is usually the case. Nevertheless, the pipette method is generally considered the best available mechanical method.

In making a mineral analysis of a sample, it becomes clear that the different size ranges in samples usually contain discrete groups of minerals. For example, sands contain a set of minerals that are most often completely different from those within the size range of clays. Therefore, in comparative analysis, it is important to make comparisons within the same size range. The results of comparing minerals found in one size range in one sample with a different size range in another sample would be deceptive.

When a transfer of soil takes place between two objects in a place, the transferred sample seldom truly represents the distribution of sizes in the soil at that place. Thus, grain size distribution in two samples from the same place may not compare exactly. For this reason, analysis of particle size distribution can contribute to the conclusion of comparison or lack of comparison, but it is seldom definitive in itself.

Stereo Binocular Microscope

The stereo binocular microscope is extremely useful to the forensic geologist. The information it allows him or her to collect in an examination of soil makes it the next logical step after analyzing color.

Light microscopes are generally of two types: transmitted light and reflected light. In transmitted light microscopes, the light source is beneath the specimen, which must be transparent. Biological microscopes used in studying tissue are of this type. In reflected light microscopes, the light source is above the object, allowing its surface features to be viewed. Such a microscope is essentially a stationary, higher-power magnifying glass. Most of these microscopes have two sets of lenses, and thus the object is viewed in three dimensions—that is, in stereo. To find the magnification of a microscope, multiply the magnification of the ocular lens, which is commonly 10X, by the magnification of the objective lens. This latter differs from microscope to microscope but is seldom more than 10X. This gives a maximum magnification of approximately 100X. Objectives may be individual lenses of fixed magnification, or some microscopes use a zoom objective that can change magnification continuously from less than 1X to about 5X. Most viewing with stereo binocular microscopes occurs at magnifications of 10X to 40X. Some of these microscopes have a second light source in the base so that objects can be viewed in both transmitted and reflected light. When the transmitted light is polarized, the microscope may be used for both stereo reflected light viewing and low-power transmitted polarizing light studies.

Stereo binocular microscopes allow examiners to view objects as small as approximately 10 microns in diameter. The upper limit is determined by how large a sample can fit under the instrument so that the surface of pebbles and cobbles can easily be viewed. Light-colored minerals are usually placed on a tray with a dull black finish for easier viewing, and dark-colored minerals get a tray with a white finish. Various inserts available for these microscopes permit the measurement of object size or provide grids for counting particles. Sample trays sometimes have etched grids for the same purpose. Trays are also available with gummed surfaces to hold grains in place for easier counting.

In examining a soil sample or similar material, the scientist first examines the whole sample as it is received, observes the types of grains and particles, and records a general impression of the material. Nonmineral materials such as metals, hair, fibers, paint, and plastic, which could be extremely valuable evidence, are removed for further examination by specialists. As we have seen earlier, plant particles can be of great value. The amount of plant material in a sample is usually less important for forensic purposes than identification of individual grasses, seeds, leaves, and the like.

The following case provides an example of the importance of nonmineral materials, especially plants. Following a bank holdup the suspects abandoned their getaway vehicle and picked up another car that had been parked in a rural wooded area adjacent to a field. They were apprehended in the second vehicle but refused to reveal the location of the original getaway car. Examiners looked at soil removed from the frame and fenders of the second car. There was a great deal of soil, indicating that the car had been driven off pavement through moist soil. Binocular microscopic examination of the soil revealed hair from both brown and black cows and a palomino horse. Seeds, leaves, and other particles of vegetation made it possible to reconstruct the types of plants growing where the car picked up the soil. Examination of the minerals revealed the soil was formed by the weathering of limestone, indicating that the area in question must be underlain by limestone rock. Forensic geologists determined the limestone was a particular type of rock underlying several square miles in the area. Armed with this information, investigators drove the back roads of the area covering the limestone, looking for a field with particular vegetation and the right animals. They found it. The missing car was in the woods on the edge of the field. Combined with evidence from the second suspect car, discovery of the getaway car helped in the reconstruction of the crime and conviction of the suspects.

In another clear case of the value of plant material, in a suburb of Sydney, Australia, in 1960, the eight-year-old son of a recent lottery winner was kidnapped and murdered. Six weeks after the kidnapping, boys playing in a wooded area found the victim's body. Examination of his clothing under the binocular microscope revealed, among other bits of evidence, pink chips composed of grains of sand in a matrix that reacted with hydrochloric acid: the mineral calcite. These chips turned out to be pieces of mortar used in house construction in the area—especially the construction of one-family houses on high foundations. In examination of plant material associated with the body, botanists found abundant parts of two cypresses, *Chamaecyparis pisifera 'Squarrosa,'* and *Cupressus glabra.* The first is fairly common as an ornamental plant. The second is rather rare. Samples from the scene where the body was found failed to turn up similar plant material or particles of mortar. Investigators reasoned that the victim had first been taken to a house with pink mortar and at least two varieties of cypress trees. Carrying branches from the two types of cypress, investigators scoured the area for weeks. They found several houses with pink mortar, but no cypresses. Finally, a house with all three variables was found. The occupants were new; investigation revealed that the former occupant, a prime suspect, had moved out on the day of the murder. He was identified and convicted.

Preliminary examination of a whole sample with the binocular microscope is normally very difficult. The mixture of particles of all sizes obscures the grains and makes identification difficult. The presence of organic material contributes to the problem. For clear study of minerals and rock, the sample must be cleaned. Sieving removes the larger particles and organic fragments. If the sample is carefully washed in water, the lighter organic particles will generally float and can be removed and saved for study. Treatment with hydrogen peroxide removes fine organic matter. The use of ultrasonic cleaners can change and damage a sample. A sample that had chips of red shale before ultrasonic cleaning may contain thousands of silt-sized quartz grains and clay minerals—the shale broken into its components—afterwards. If the examiner is certain that various rocks and minerals are not affected by ultrasonic cleaning, then the method can be useful.

Using a stereo binocular microscope, the experienced scientist can identify the rocks and minerals in a clean sample on sight or through simple tests. It is possible to observe the texture and coatings on the surface of the grains, and such properties as shape, rounding, weathering, inclusions, color, and polish. The counting of different kinds of grains is especially important. Recorded numbers are normally more useful than qualitative impressions. However, the samples may be so different on first examination that further work is not useful because a determination of comparison could never be made. Microscopic determination of the number or percentage of different types of grains is extremely important in determining comparison or lack of comparison. In counting grains of different types, it is important that the sample be representative of the whole sample; that identification be consistent and accurate; and that the same grain not be counted twice because the sample moved. Scientists can usually avoid the last problem by placing the sample on a gummed surface or by removing grains as they are counted and placing each, as removed, in a container or gummed individual tray. The scientist's judgment and caution are the most important factors, whatever the method used.

While much can be learned about minerals through examination with the stereo binocular microscope, further study of rock in thin section may be required. In a substitution case, wooden boxes filled with gold bars were

Microscope laboratory showing a stereo binocular microscope
—COURTESY OF MCCRONE ASSOCIATES

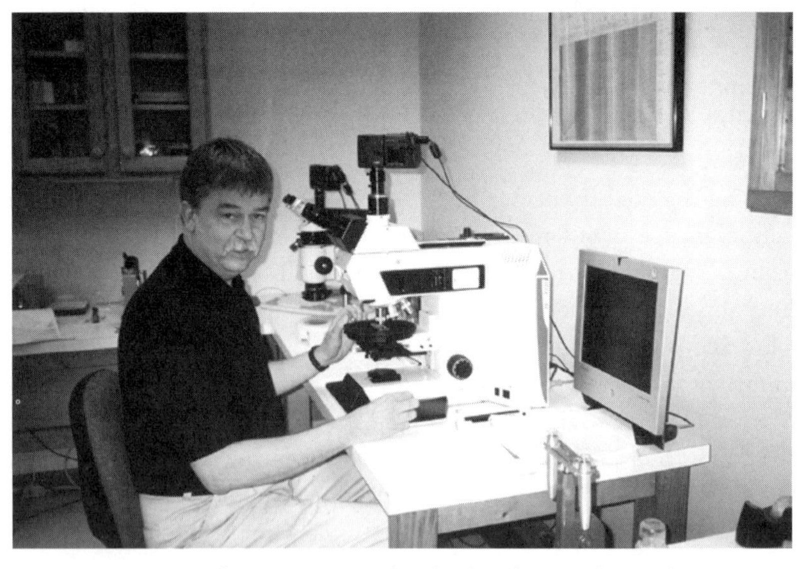

Petrographic microscope used in the identification of minerals

flown to England. When the boxes were opened, they contained nothing but stream-worn pebbles. For insurance purposes, the owner, the insurance company, and the airline needed to know where the transfer took place. Examination of the pebbles showed that they were derived from rocks found in the Alps and were common in certain rivers that flow from the Alps to the sea. The plane had stopped in Milan, Italy. Forensic geology established that the transfer took place there, prior to the shipment's arrival in England.

Petrographic Microscope

Petrographic microscopes differ in detail from ordinary compound microscopes. However, their primary function is the same: to produce an enlarged image of the object that is on the microscope stage. The combination of two sets of lenses, the objective and the ocular, produces the magnification. The function of the objective lens—at the lower end of the microscope tube—is to produce a sharp, clear image. The ocular lens merely enlarges this image. For mineralogical work, examiners generally use three objectives: low-, medium-, and high-powered. The magnification produced by objectives is usually 2x (low), 10x (medium), and 50x (high). Oculars have different magnifications, usually 5x, 7x, 10x, 15x, and 20x. Total magnification of the image is determined by multiplying the magnification of the objective lens by that of the ocular lens as follows: 50x times 10x = 500x.

Ocular lenses normally contain a crosshair that helps locate grains under high power when changing objectives. A condensing lens system is normally provided under the stage for use with high magnification and to help view the various optical effects that minerals produce. Petrographic microscopes have a rotating stage and, under the stage, a polarizing filter that transmits light vibrating usually in a N-S (north-south, or front to back) direction. Above the stage, in the tube of the microscope, is a second, removable, polarizing filter. It transmits light usually in an E-W (east-west, or right to left) direction. When the upper filter is inserted while the lower one is in place, it blocks light from passing through the microscope. In this case the filters, called polars, are said to be crossed. Only when an anisotropic material (a material that is not isotropic—i.e., it is formed in the isometric crystal system or is amorphous) is placed on the microscope stage under crossed polars can it be seen. The anisotropic mineral rotates the N–S vibrating light from the lower polarizing filter, permitting some of it to pass the upper E–W polarizing filter. When the stage is rotated, there will be four positions where the vibration directions in an anisotropic crystal will line up with the N-S and E-W direction. At these, the crystal is said to be "at extinction": it appears black and is thus invisible.

In identifying mineral grains under the petrographic microscope, it is common to use the immersion method. Mineral grains are placed on a microscope slide in a liquid of known refractive index (available commercially). The range of liquid refractive index of 1.46 to 1.62, with a difference of 0.02 between adjacent liquids, serves most purposes. When the grain is viewed, a narrow line of light commonly surrounds the grain. If the distance between the objective and the sample is increased slightly, usually by raising the tube of the microscope, the line of light, called a Becke line, moves in the direction of the higher refractive index. If the mineral has a higher refractive index, the Becke line moves into the grain. If the liquid's index is higher, the Becke line moves away from the grain into the liquid. By trial and error with different liquids, a match is found, by which point the grain is almost invisible in the liquid. In most cases the grain's refractive index falls between the refractive indexes of two liquids, and the experienced observer can estimate value.

Consider a mineral in the isometric crystal system. Since it is isotropic—that is, it has only one refractive index—it remains dark at all positions under crossed polars. When the upper polarizing filter is removed, the examiner can see the mineral and determine its refractive index. Knowing the refractive index and the fact that the mineral is in the isometric crystal system, together with observations of color, cleavage, and the like, the microscopist can identify the mineral.

It is easy to obtain information about minerals that form in other crystal systems. Hexagonal and tetragonal minerals have two refractive indexes. It is also possible to find out whether the mineral is positive or negative. To do this, the scientist inserts accessories into the microscope and finds which of the two refractive indexes is higher. Identification depends on the two indexes of refraction, the optical sign (positive or negative), and other properties. Several books list the optical properties that facilitate identification (for example, McCrone and Delly, 1973).

The petrographic microscope, an important tool in many aspects of forensic work, is the best tool for studying the optical properties of rocks and minerals. The study of individual mineral grains or thin sections of rocks and related material is easily accomplished by anyone trained in use of the instrument. For a thin section of rock, the rock is cut with a diamond saw and the surface of the slice polished. This polished surface is cemented to a glass microscope slide with an adhesive of known refractive index such as epoxy or Canada balsam. The scientist then makes a saw cut parallel to the glass, leaving a wafer of rock cemented on the slide. Grinding of the wafer proceeds to a thinness of approximately 30 microns. A thin glass cover is then glued to the polished rock surface to protect the rock and improve viewing. Most rocks are transparent at this thickness and can be viewed in transmitted light. Similarly, loose mineral grains of the same general size are commonly mounted in epoxy or Canada balsam on a microscope slide and covered with glass for microscope study. This is the method used when heavy minerals—minerals with high specific gravity, such as rutile, garnet, zircon, and tourmaline—are separated from common lighter minerals, such as quartz and feldspar, through settling in sodium polytungstate or one of the other tungsten-based heavy liquids.

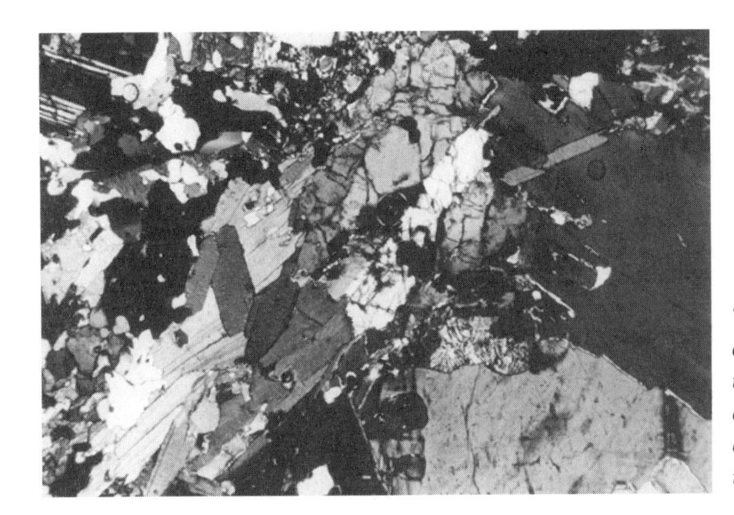

Thin section of basalt (an igneous rock) as seen through a petrographic microscope

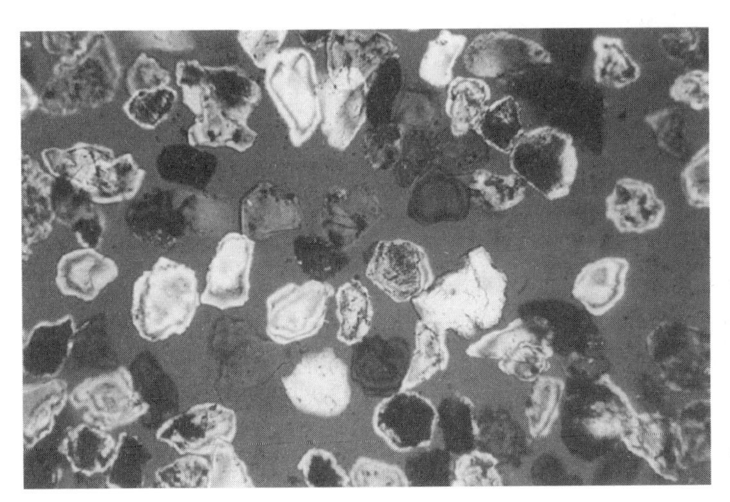

Heavy mineral grains mounted on a glass slide and viewed through a petrographic microscope

In a well-known case, study with the petrographic microscope led to the conviction of a thief of sheep wool. A suspect in possession of a number of fleeces in Wyoming claimed they came from sheep far from the scene of the theft. The crime scene was underlain by isolated outcrops of weathered red shale and sandstone of the Permian Satanka Formation. Examination by petrographic microscope of the dust on the wool indicated that the minerals compared with those in the area of the red shales and sandstones and were also different from those in the area from which the suspect claimed the fleeces had come.

The petrographic microscope is useful for identifying materials other than minerals and glass. These include starch grains, synthetic crystals such as abrasives, cements, and ceramic materials. In a murder case, the body of the victim was found in a shed normally used for storing garbage. The floor was covered with smashed potatoes and potato skins. A suspect was found whose shoes were covered with starch. While this would seem to match details of the case—anyone walking on the shed floor would acquire starch on his shoes from the potatoes—examination with a polarizing microscope determined that the starch on the shoes came from wheat, not potatoes. The suspect happened to work in a bakery, where the floor was covered with wheat flour. In this case first impressions did not prove to be true.

Identification of Rocks in Thin Sections

Information gleaned from the study of thin sections is often useful in identifying rock and comparing samples. A case in point involved the Pennsylvania Geological Survey. The Industrial Development Department of the Penn Central Transportation Company submitted two rock samples to the survey associated with the following problem. New automobiles from Detroit shipped to New Jersey by train were arriving with smashed windows, dents, and damage resulting from rocks thrown at the passing railroad cars. Could examination of the rocks found in the automobiles help identify where the vandalism was taking place so police procedures could be initiated? Obviously, the whole length of track between Michigan and New Jersey could not be policed. In addition, the automobiles were transported along two routes, one through New York State and one through Pennsylvania. Along both routes were an incredible variety of rock types, some of which occurred at several different places along the same route. Thin sections of two submitted samples were examined microscopically in the Pennsylvania Geological Survey laboratory. Both rocks were found to be coarse-grained (pegmatitic) gneiss containing feldspar, quartz, biotite mica, chlorite, and slender crystals, probably of the mineral apatite. These minerals and the rock texture provided the critical clue that the rock specimens came from a metamorphic terrain.

Investigators narrowed their search to two possible areas, southeastern New York and eastern Pennsylvania. These geologic areas—the Reading Prong in New York and the Piedmont in Pennsylvania—both contain metamorphic rocks. The rock type of the samples recovered from the damaged vehicles occurs along the Penn Central Railroad in Pennsylvania, but it usually has less biotite and seldom any apatite. On the other hand, rocks containing these minerals are common in a limited area of southeastern New York State. The survey suggested that the most likely source of the thrown rocks was along a stretch of tracks in the vicinity of West Point, north of New York City. After Penn Central's own geologists confirmed this in an independent study of the northern route, Penn Central Railroad police concentrated on the West Point area. Several of the culprits were spotted doing the damage and apprehended.

Heavy Minerals

Heavy minerals generally have a specific gravity greater than 2.89. They usually represent only a small part of a soil sample, but they can be very useful for characterizing the material. For this reason, heavy minerals have long been used in geologic studies that attempt to recognize similar rocks and to determine the kinds of rocks that were weathered to produce the particles for sedimentary rocks. Prior to separating the heavy minerals in a sample with heavy liquids, the sample is fractionated into size ranges. Various size ranges can be selected for study, but the size range between 0.5 and 0.1 millimeters is commonly used. Concentrated by settling in the heavy liquid, the minerals are then transferred to a microscope slide covered with a mounting medium of known refractive index, such as Meltmount, Norland Optical cement, Lakeside 70 or Canada balsam. Then they are studied with the polarizing microscope. Individual mineral grains are identified and counted, and results generally reported as each mineral's percentage of the total makeup. Opaque grains such as magnetite and ilmenite are usually grouped together and identified, after surface polishing, under the kind of reflecting polarizing microscope used in metallography, where polished surfaces of metals are examined.

Examination of heavy minerals had important evidential value in a New England murder case. The victim's body was found on a beach, and the suspect had sand in his shoes. He said he had picked it up by walking on nearby beaches and that he had never walked on the beach associated with the crime. Studies of the heavy minerals in sand from the crime scene and from nearby beaches demonstrated that the minerals in the shoes were similar to those from the beach where the body was found and were not similar to the sands of the other beaches. The most helpful mineral diagnostically in this case was the rather rare black tourmaline schorlite, which varies widely in abundance from location to location.

In January 1968, the body of a man named Koklas was discovered in Australia near the Barkly Highway, apparently a victim of murder. A rather complicated set of circumstances led to the arrest in Perth of a suspect named Da Costa. The suspect admitted to having traveled with Koklas from Melbourne to Mount Isa, but claimed to have argued with and left him there, taking some of his possessions. These included a pair of bloodstained shorts, with sand adhering to the bloodstains. Mount Isa is over 300 miles east of where the body was discovered, but it was suspected that the shorts had been removed from the body where it lay. Experts examined the heavy minerals in samples of the sand on the shorts and samples from the scene of the crime. The sand on the shorts was consistent with that from the crime scene, with slight contamination. Additional samples from the area helped determine the degree of variation in the heavy minerals there. In particular, grains of tourmaline in the samples were analyzed with an electron probe microanalyzer. Variations in the heavy minerals and the tourmaline compositions showed that the samples closer to the scene compared more closely with the sample from the scene, and it could thus be established that the sand on the shorts was much more likely to have come from the scene of the crime than elsewhere in the area, or from Mount Isa. After a lower-court hearing, the suspect admitted having been at the scene of the crime.

Refractive Index

A transparent material's index of refraction is the ratio of the velocity of light in a vacuum—normally considered to be 1—to the velocity of light in the material being analyzed. Thus a refractive index of 2.4553 means that light travels 2.4553 times as fast in a vacuum as in the transparent material. The measurement of refractive index, which is one of the most important methods for the comparison of glass, may be made using the Becke line method, as previously discussed for minerals. Performed by a competent scientist, this method produces results accurate to plus or minus 0.003. However, more accurate results can be obtained with less effort using either the single- or double-variation method. In the single-variation method, a chip of glass is placed in a liquid of known refractive index on the microscope slide. The liquid should be one whose refractive index changes in response to temperature. The microscope stage has a heating element that slowly warms the slide. As the liquid warms, its index changes. When the liquid and the glass match, the examiner observes and records the temperature. Tables are available of the refractive index of these liquids at different temperatures. The temperature determines the refractive index of the liquid and the glass. It is possible to check the refractive index of the liquid at any temperature using glasses of accurately known refractive index.

Refractive index liquids show dispersion; that is, they have different refractive indexes in various colored lights. Thus the color of the light may be varied until a match is produced and the refractive index of the glass determined. This is possible because glass, like most other solids, shows little dispersion compared to liquids. Combining the two methods of temperature and color is known as double variation. The standard refractive index is obtained with orange light (5893-angstrom wavelength), and it is this value (sodium D at 20 degrees Celsius [68 degrees Fahrenheit]) that is commonly reported. With these methods, accuracy greater than plus or minus 0.001 can be obtained. Variation in the color of light permits the direct measurement of dispersion—that is, the difference in refractive index obtained by measuring with blue light (4861 angstroms) and red light (6563 angstroms). Indexes of refraction and dispersion are the two most used properties in the identification of glass.

One new problem in identifying glass is that quality control has, in recent years, reduced the range of refractive index in common glass. Small variation still exists within any single plate of glass. Thus, to be forensically useful, many laboratories use a phase contrast microscope to measure the index. Phase contrast results from additions to the microscope. With a Mettler temperature control attached to a heating microscope stage, it is theoretically possible to detect refractive index differences as small as 0.00004. This is possible because many refractive index liquids drop in index 0.0004 for each degree Celsius the liquid is heated. The Mettler heating stage can be controlled to 0.1 degrees Celsius. The examiner then records the temperature at which a refractive index match occurs, noting whether the temperatures and wavelengths of light are similar for two samples to determine whether they compare. Variations in refractive index within a single piece of glass are important in determining whether

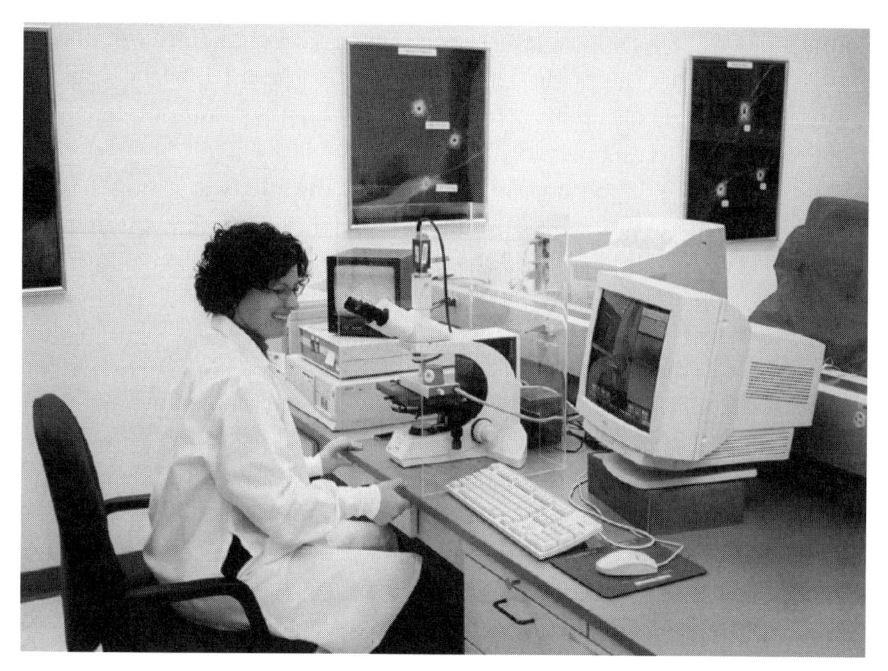

GRIM II instrumentation for the determination of glass refractive indices

questioned and known samples compare. A single piece of plate glass can have refractive index differences of 0.0001.

The relative specific gravity of glass fragment samples can be obtained using a heavy liquid. Two fragments of glass are placed in a tube with a heavy liquid such as bromoform. This liquid is diluted with a lighter liquid such as ethyl alcohol. The glass fragments sink if they are denser than the liquid. If they sink, more of the heavy liquid is added until they float freely, neither rising nor falling. If both fragments float in this way, they have the same specific gravity. If one very slowly rises or sinks in relation to the other, they do not have the same specific gravity but are close enough to be within the variation found in a single large pane of glass, 0.0003 grams per cubic centimeter. It is important that the two liquids be thoroughly mixed and maintained at a constant temperature. In this method, actual specific gravity is not determined, but that is not important since the two glass samples are directly compared. In order to measure specific gravity directly many laboratories use a density meter to determine the density of a fragment. These instruments produce number values and permit the measurement of different fragments at different times.

Other properties useful in comparing glasses are color in white light and ultraviolet light; chemical composition; and thickness. It must be remembered that the thickness of glass is almost never uniform. This is especially true for hand-blown glass. In some cases it is possible to mechanically fit chips of glass back together and to match the characteristic markings on the broken surface, resulting in a very high level of confidence in the comparison and truly individualized evidence.

Glass fragments were helpful in confirming the identities of two burglary suspects. At four o'clock in the morning, just down the road from Stanford University, police responded to an alarm at a sporting goods store. When they arrived, they spotted two young men driving away in an older-model car. The police gave chase and captured the two men. Examination of the car revealed several very small fragments of glass stuck in the front bumper and fender. Asked for an explanation, the suspects suggested that someone backed into them while parking. Back at the sporting goods store, investigators discovered that forced entry had been made by ramming a car against the store's rear door. Glass fragments lay on the ground both inside and outside the door. Laboratory examination of the glass from the car bumper and from the sporting goods store back door by refractive index, density, color, and thickness showed them to be identical. The two young men pleaded guilty.

Cathodoluminescence

A luminoscope, attached as a stage on a microscope or a scanning electron microscope, is used for measuring cathodoluminescence. The specimen—mineral grains, for example, or a thin section—is bombarded with a beam of electrons generated by the instrument. Electrons striking the surface of the specimen produce an optical luminescence seen as a display of colors. The colors and their intensity depend in large part on very small changes in the concentration of trace impurities, the minerals present, and the location of the trace impurities in the structure of the minerals. The method has wide application in determining and observing a variety of differences in mineral grains that otherwise appear similar.

Scanning Electron Microscope

The scanning electron microscope (SEM) has a wide range of magnifications, generally from 25× to over 650,000×, and can record something as small as 1.5 nanometers. Needless to say, most of the forensic geologist's work falls well within these limits. The SEM became commercially available in the mid-1960s and was rapidly introduced into forensic work, especially in the study of gunshot residues and various other very small particles. The SEM has the advantage of allowing direct viewing of the surface of a sample. However, an ultrathin coating of carbon or gold on the specimen improves the quality of the picture. The depth of field is very large, and most SEM pictures have an excellent three-dimensional appearance. The examiner can change magnification easily and thus study surfaces with very low to very high magnification. Previously undetected differences in very small fossils can now be seen during routine examination. The SEM shows scratches, pitting, and mineral growth on the surfaces of individual grains of minerals such as quartz. These features may prove useful in telling us the past history of the individual grain. The SEM can even reveal other minerals, such as clay flakes filling scratches, adding another possibly useful characteristic in comparing minerals.

Wayne Isphording has worked many civil cases involving geology and is known for his creativity in approaching problems. Use of the scanning electron microscope was critical in one case. On November 7, 1980, a motorcyclist entered a curve in a fog-shrouded, marshy area of southern Alabama. Later, he stated that his helmet visor suddenly fogged up and that he could not raise it. The motorcycle slid out of control, throwing the rider into the path of an oncoming car. The accident ultimately resulted in surgical removal of the rider's left leg. The victim sued the local motorcycle dealer and the national motorcycle company, alleging that he had not been told that the visor, obviously secured with four large rigid snaps, could not be raised. A number of lines of evidence argued against the plaintiff's version of events. The local dealer denied that the helmet was one it carried or sold. Index of refraction, X-ray diffraction, and fluorescence chemical analyses supported their claim, clearly showing that the visor on the plaintiff's helmet was different from the visors sold by the defendant. Further, although the plaintiff claimed he had purchased the visor the night before the accident, the visor had a curvature possible only after weeks of attachment to the helmet. Most interestingly, SEM examination of the inside of the plastic visor disclosed scratches that contained very small grains of orthoclase feldspar. The accident site, however, was on the Plio-Pleistocene Citronelle Formation, which is completely devoid of feldspar. The mineral inside the helmet must have been acquired before the accident in question and at least 150 miles away. Each of these points helped build a case that the visor had not been purchased the night before the accident at the defendant's store. After viewing this evidence, the attorneys for the plaintiff removed the local dealer from the suit.

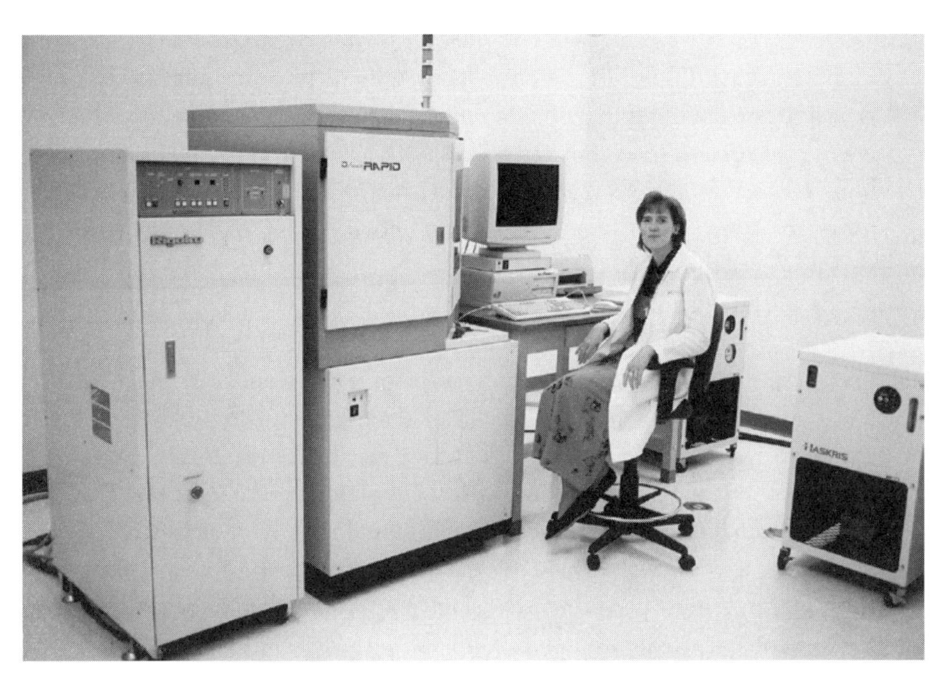

X-ray diffraction instrumentation for identification of minerals

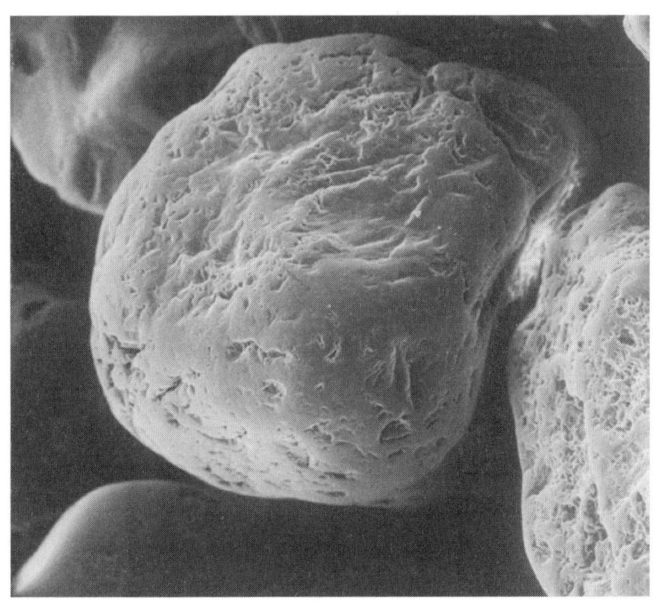

Quartz sand grain seen with a scanning electron microscope at approximately three hundred times magnification

When using the SEM in forensic work, with its high magnifications, it is well to keep in mind that no two objects are ever exactly the same. This is even true of two sand grains that have been side by side for the past million years. Observations made with these instruments can be very useful for establishing similarity or dissimilarity between samples. However, the very power of the instruments permits the possibility of their abuse in the hands of the unskilled or unscrupulous. Complete chemical analysis of a person by the most modern methods in the morning and repeated in the afternoon would show chemical differences. It would not mean we had analyzed two different people. Similarly, the demonstration of small differences in soil does not prove that they do not compare. It is equally true that showing a similarity among soil samples—for example, that they both contain quartz, the most common mineral in soil and sediment—is poor evidence on which to base a comparison. Thus the professional judgment of the scientist becomes increasingly important when these powerful instruments are used.

Scanning electron microscopes can determine the elemental composition of the particles being examined. X-rays are produced when the electron beam of the microscope strikes a target. The SEM can be coupled to an X-ray analyzer, which sorts the energy or wavelength values of the X-rays. Since these are related to specific elements, they shed light on which elements are present in the material being viewed. In addition, the intensity of the emitted X-rays correlates with the amount of each element present. Thus the examiner can determine the chemical composition of individual particles.

X-ray Diffraction

X-ray diffraction is one of the most important and reliable methods of identifying the composition of crystalline substances. The method focuses on the arrangement of atoms, ions, and molecules within the specimen. X-rays are passed through crystals and the angle of the diffracted rays measured. Each crystalline material has its own distinctive X-ray pattern. The X-ray diffraction pattern of a sample is controlled by the sample's internal structure. The pattern can be collected on film, on an image plate, or with an electronic detector. Under normal circumstances, interpreting X-ray patterns is a simple matter.

There are at least two ways to interpret X-ray diffraction data. In the first, d-values and intensities are measured and compared with published lists of data on minerals. In the second, the X-ray pattern is compared directly with the pattern of a known mineral. Sometimes the X-ray diffractograms of two samples may be compared without actual identification of substances, but this is less useful as evidence than actual identification.

One of the strong points of X-ray diffraction is that the patterns record crystal structure. For example, analyzed chemically, diamond and graphite would seem to be identical since both are composed of pure carbon. However, X-ray diffractograms of the two minerals would reveal that they are quite different. Many samples are mixtures of two or more substances. In addition, with chemical analysis, the actual chemical form of the substances cannot always be established. As an example, we can use a mixture of two salts, sodium chloride and potassium nitrate. The usual chemical methods of analysis would reveal a composition of sodium, potassium, chloride, and nitrate. But what were the original compounds? Sodium chloride and potassium nitrate? Sodium nitrate and potassium chloride? Or a mixture of four salts? An X-ray diffractogram of this salt mixture would tell us the specific form of the salt.

X-ray diffraction is the principal tool in modern identification of clay minerals. The chemical composition of clays generally tells us very little about their nature, but the possibilities of identifying clays by X-ray diffraction are almost unlimited. Clays, like other crystalline substances, can be X-rayed and identified with the help of standard reference books or cards.

X-ray diffraction methods provide a definitive identification of minerals and other crystalline substances. However, when a sample contains several mineral kinds, the results can be confusing. It may be impossible to identify minerals that exist in small amounts or the amount of any given mineral in the sample. This is generally true because determination of the amount is influenced by the orientation of the grains on the sample holder, their crystallinity, and their inherent properties. In many cases, an examiner has simply X-rayed a total soil sample or pieces of concrete and used comparison of the patterns as evidence. But generally such results are impossible to interpret and, like any bulk method, it does not use the basic value of forensic geology, which lies in the vast diversity of minerals. When most soil samples are X-rayed, the patterns appear to be very similar or virtually identical. This is because quartz and to some extent feldspar usually make up the bulk of the material. In such instances it would be more definitive to isolate the heavy minerals or use some other form of sorting, grind the resulting minerals to a powder, and make diffractograms for comparison. Considerable research is being conducted in an effort to produce reliable methods for quantitative determination of the amount and kinds of minerals in a bulk sample. When this is accomplished and shown to be reproducible, we will have a most valuable tool for the forensic examination of earth materials.

X-ray diffraction analysis is also used in the identification of explosives that contain inorganic crystalline material such as potassium nitrate and potassium chlorate.

QEMSCAN

QEMSCAN, an automated scanning electron microscope system, analyzes the mineralogy of inorganic deposits. QEMSCAN stands for Quantitative Evaluation of Minerals by Scanning electron microscopy. Samples submitted for analysis are embedded within resin and allowed to harden, and then the solid blocks are polished. As the electron beam moves across a particle embedded in the resin, the energy X-ray spectrum is measured at an operator-defined pixel spacing. The spectrum of each pixel is then compared against a database of known spectra developed by CSIRO Minerals in Australia and assigned a mineral or phase name. Depending upon the predefined pixel spacing and the particle size, QEMSCAN can quantify the composition of approximately one thousand 1- to 10-micrometer-sized particles per hour. Once the pixel spacing and mode of operation have been selected by the operator, data collection is entirely automated. If a spectrum is measured that cannot be identified, the operator can manually reexamine the composition of the unidentified pixels. In addition, the system can measure and identify many noncrystalline phases that have distinct elemental compositions such as fly ash, glass, or ceramic products. However, the system cannot identify organic material. Because the phase analysis is based upon the acquisition of X-ray energy spectra, different phases that have the same chemical composition, such as diamond and graphite, cannot be separately classified. QEMSCAN moves us a big step toward the goal of rapid quantitative mineral analysis for forensic purposes.

Paramilitary groups involved in the conflict in Northern Ireland have been on ceasefire since the late 1990s, but some still maintain criminal activity, including property speculation, armed robbery, extortion, illegal gaming machines, and drug importation and dealing. These loyalist groups increasingly have found themselves operating against one another in the sort of urban gang warfare common to other major cities of the world. In 1997 two such gangs controlled East Belfast. In May, a hoax telephone call forced one gang leader to open the gates of his well-fortified country house, whereupon two gunmen hired by the rival gang entered his driveway and shot him repeatedly in the head. Alastair Ruffell, a distinguished forensic geologist in Belfast, undertook the case.

Following the fatal shooting, the gunmen were unable to use their getaway car because neighbors appeared, alerted to the noise. Thus the assailants moved from their firing position into an adjacent property, climbed a locked gate by jumping on the top of a car, and landed in a yard with various waste materials in it, including broken bottles, broken plaster board, a leaking heating oil tank, and abundant leaf litter. Some evidence of the assailants falling or slipping in this location was subsequently noted at the scene by crime officers. From this point onwards, a police tracker dog and handler were able to follow the trail of the assailants, finding where they disposed of some outer clothing and one of their weapons in the garden of the same house and made their way into the adjacent forest and farmland. A barbed-wire fence along the escape route had fresh blood on it. The dog lost the scent of the escapees at a lane some distance from the scene where it is supposed another getaway car had made a successful pickup, but tire tracks were indistinct following heavy rain showers. Police intelligence revealed a number of likely suspects in the shooting, one of whom was arrested. This suspect had abrasions to his body and lacerations to his arm. His blood matched the DNA of the blood recovered from the barbed-wire fence, but this evidence only placed the suspect some few hundred yards from the scene. Also recovered at the time of arrest was the suspect's vehicle, which was generally clean inside and out, except for some cream-to-white powder and fragments on and in the tread of a rubber foot mat. Preliminary geologic examination revealed this to be gypsum with associated plant fragments and some oil. It is presumed that the suspect stepped on the gypsum board while escaping and then transferred the material from his shoe to the floorboard of the escape vehicle. The total amount of recovered material was less than 0.5 milligrams. This very small amount of material is entirely typical of urban trace evidence as well as in cases where the suspect is well-versed in cleaning-up operations. Analysis for comparison with the plaster board and associated debris found on the escape route required a method which would provide good statistical data from a small sample. Traditional geologic analysis including microscopic examination and analysis by X-ray diffraction managed to establish the presence of gypsum but failed to discriminate any differences in the samples: traces of over ten plaster-board samples were indistinguishable. For this reason, QEMSCAN was deployed. Ruffell asked Duncan Pirrie, who operates a QEMSCAN laboratory in Great Britain, for help.

QEMSCAN analysis demonstrated that despite the dominance of gypsum, the samples contained abundant minor amounts of a wide collection of minerals and chemical phases. The aim of the investigation was to test whether the samples recovered from the motor vehicle floor mats were comparable with the samples recovered from the scrap plaster in the yard on the escape route, and to consider how distinctive different sources of builders gypsum were. The particulate material recovered from the floor mat of the motor vehicle was found to be comparable mineralogically in terms of both the major, minor, and trace minerals and phases present with the plaster samples recovered from the yard. The analysis of four commercial plaster samples revealed mineralogical and in some cases textural differences between these different plasters.

The suspect, Robert Young, was brought to trial in the autumn of 2005. His involvement with paramilitary groups, the discovery of a pistol magazine in his house, the presence of fresh blood with his DNA on the escape route, and the comparison of vehicle plaster board samples with escape-route samples led to a conviction. His appeal against this decision was not sustained.

Chemical Methods

Many instruments and methods measure the chemical composition of organic and inorganic materials. They can provide valuable information. For example the chemical composition of glass is often a valuable addition to knowledge of its optical and physical properties. Identification of fertilizer in a soil adds a whole new dimension to the list of properties used in comparison. There is a classic case of stolen potatoes where chemical analysis of the soil was critical. The potatoes were stolen from storage in an eastern U.S. city. Police developed a suspect who had a large inventory of potatoes. Analysis of his potatoes showed that the soil clinging to them had a superphosphate content similar to soil from the farm where the stolen potatoes had been grown. In support of the evidence, the farm was heavily fertilized and its soil had a buildup of phosphate. The suspect was convicted.

Some methods for determining the amount and kind of elements in a sample rely on the fact that, as one of their properties, elements selectively emit or absorb light. This is the basis for emission spectroscopy and atomic absorption spectrophotometry. Neutron activation analysis is a nondestructive method with a detection sensitivity of one-billionth of a gram. In this method, a nuclear reactor bombards the sample with neutrons. The resultant gamma ray radioactivity is measured to identify the presence and amounts of elements. Needless to say, this method is expensive to operate and maintain and leaves the samples radioactive. Organic compounds contain carbon; their identification requires different methods.

There are several techniques for separating out the various compounds in a mixture. Generally these methods depend on the relative amounts of the gas phase of a compound and its liquid phase under fixed conditions. The amounts are characteristic properties of each compound. Because those compounds that have a higher tendency to go to the gas phase move more quickly, the distance they move in a given time identifies and separates the compounds. Generally, these methods are called chromatography. Scientists also identify compounds with spectrophotometers that measure light-absorbing properties. Mass spectrometers have the ability to uniquely identify compounds if analysis is done under proper laboratory procedures. The mass spectrometer bombards the sample with high-energy electrons, causing molecules to lose electrons and become positively charged. In this unstable state, the molecules break into fragments. The instrument then passes the fragments through an electric or magnetic field where they are separated according to their masses. This permits specific identification of the compounds, since distribution of masses is a unique property.

Fourier Transform Infrared Spectroscopy and Raman Spectroscopy

Fourier transform infrared (FTIR) spectroscopy and Raman spectroscopy are nondestructive analytical tools commonly applied to the identification of organic minerals and material. Often combined with a microscope, they allow the identification of very small objects or parts of a sample. The nondestructive aspect is important since, in the analysis of art objects or gems, most owners do not appreciate removing parts for analysis. In addition, in criminal cases, the sample may be very small. Removal of material for analysis may leave little of the sample for additional study or verification. In these methods, a source generates light across the spectrum of interest. The sample absorbs light according to its chemical properties. A detector collects the radiation that passes through the sample. Computer software analyzes the data collected and compares results with known spectra of organic and inorganic materials. Professor Howell G. M. Edwards of the University of Bradford in Great Britain has reported on a case in which a beautifully carved cat was long thought to be ivory and about 300 years

Raman spectrometer for nondestructive analysis of materials

old. The aforementioned methods revealed that the cat was a fake: It was actually composed of polymethyl methacrylate, polystyrene, and the mineral calcite.

Density Gradient Column

The density of a mineral, rock, or other solid particle—that is, the weight of the particle per unit volume—is usually expressed as grams per cubic centimeter (g/cm^3). Density differs for each particle depending on the minerals present and the particle's chemical composition. It also depends on the amount of pore space between mineral grains and whether bubbles (fluid inclusions) exist within the minerals. The density of individual common mineral particles varies widely from almost 20 g/cm^3 for gold to only 1.7 g/cm^3 for carnallite. Some particles found in soils, particularly those of an organic nature, have a density less than that of water. Most organic particles are assumed to have a density approximating 0.9, that of water. The density of a material when compared with the density of water at 4 degrees Celsius (1.00) is called the specific gravity and is usually expressed as the number of times the material is denser than water. For example, pure quartz has a specific gravity of 2.65. Because the density of individual particles varies, the distribution of densities of various particles in a soil sample can be used for determining whether one sample of soil is similar to another. In the past, the density gradient column technique was widely used for forensic examination of soil. In many cases, it was the only method used to determine comparison. Examiners testified in court relying only on what they had seen in two columns.

For this method, two dried samples of soil are first carefully pulverized with a rubber tool, then placed on a nest of sieves and separated into different-sized fractions. Only those sizes that exist both in the control sample and the questioned sample should be examined. The smaller of the two samples, usually the questioned sample, is weighed; an equal amount from the control sample is also weighed. If the sample weighs more than 75 milligrams, a tube with an inside diameter larger than 10 millimeters is necessary. It is especially important that the two samples be of the same weight, because it is the concentration of densities that is to be studied. The columns are glass tubes usually 12 to 18 inches in length and sealed at the bottom. The tubes are placed in a rack and filled with liquids of different densities. The heaviest liquid is placed in the tube first, followed by liquids of decreasing densities, usually ten in all. The columns are then allowed to stand upright until the liquids have had a chance to mix by diffusion, producing a column of liquid that decreases uniformly upward in density. This usually takes twenty-four to forty-eight hours. It is extremely important that the two columns be produced in exactly the same way—that is, that the same amount of each liquid is added in exactly the same way. This is necessary because the two columns are to be compared. It is also important that the two be at the same temperature, because densities change with temperature.

Liquids used in the density column vary from one laboratory to another. However, the most common are bromoform (density 2.89) and bromobenzene (density 1.499). The two liquids are mixed in fixed amounts such as five volumes of bromobenzene to one volume of bromoform. Ten liquids of different densities, from pure bromoform to pure bromobenzene, are produced. These mixed standard liquids are placed layer by layer in the columns.

When the two columns have equilibrated and produced a uniform density gradient in the liquid, the two samples of equal weight are added, one in each column. Within a few hours the individual particles of the soils

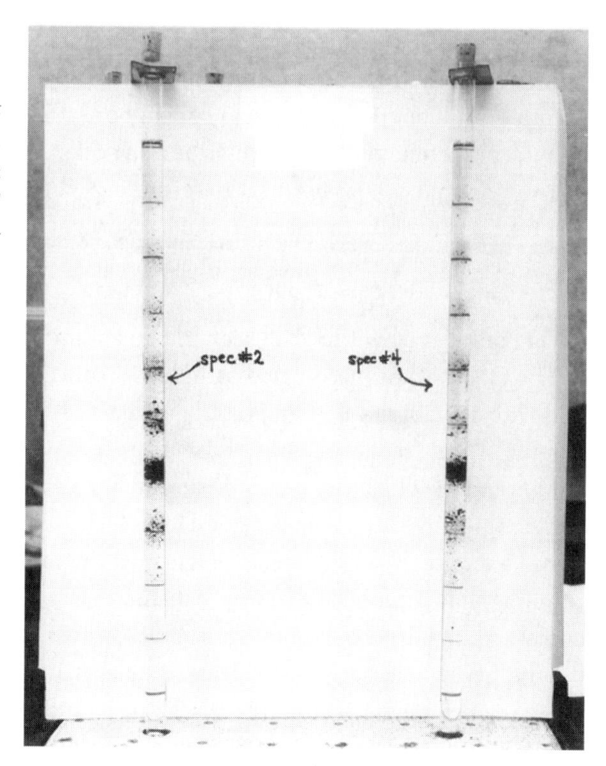

Density gradient columns. Although these two samples look similar, one is from soil adjacent to a filling station, and the other is accumulated soil from the floorboard of an automobile. —COURTESY OF NEW JERSEY STATE POLICE

settle to a level in the column where the liquid has the same density as that of the individual particle. Thus the particles will be distributed in the column according to the different densities represented in the sample. In itself, the size of the particle (except, possibly, some ultra-fine clays) has no influence on the level a particle seeks; only its density matters. Some smaller particles may take longer to come to rest. If a sample has a large proportion of fine particles, it may take up to two days to settle without further particle movement.

The distribution of particles in the two tubes is examined and usually photographed. The pictures, which may be presented as evidence, are taken against a uniform white background lighted with fluorescent or other cool light. Glass or other transparent particles may require illumination from above or below. Cool light is necessary because convection currents from hot light can disrupt the density column liquids. It is said two samples compare when the distributions of the densities of particles appear the same in the two columns. A difference of 0.01 g/cm^3 in any one segment of the density column is easily detected.

The value of the density gradient column, some argue, lies in the ease of routine comparison. The method can be standardized—that is, used in the same way in different laboratories by different people. Any skilled technician can prepare the columns, keeping human error to a minimum. But there are several geologic problems with this method, including the following.

1. Breaking up the samples with the rubber tool may produce different results in the two samples. This is especially true if one sample has been pressed and dried, as is often the case with a sample from a shoe, and the other is loose material. This can result in mineral particles adhering to each other. If, for example, a particle of mica with a density of 3.0 sticks to a particle of quartz with a density of 2.65, the composite particle will come to rest in the column at a density level somewhere between the two values. If the two particles were separate, they would come to rest at their individual density levels.
2. The density for the heaviest liquid at the base of the column, bromoform (CHBr3), is 2.89. All so-called "heavy minerals"—so helpful diagnostically in forensic soil studies—are denser than 2.89. This means that they fall to the bottom and accumulate. Another commonly used heavy liquid, tetrabromoethane (C2H2Br4), has a density of 2.97. Even with this fluid at the bottom of the column, many different minerals will accumulate. Petraco and Kubic have proposed using Clerici solution and distilled water in order to float some of the heavy minerals.
3. The most common mineral in natural soils is quartz. Quartz usually comprises more than 80 percent of individual soil samples. Pure quartz has a fixed density of 2.65. The density of an individual particle of quartz and most other minerals will change if it has fluid inclusions, solid inclusions, and particle coatings of such minerals as iron. The particles of most soils are dominated by one mineral with the same density. Differences observed in the density column result from small variations, such as inclusions or coatings, that might or might not be significant if studied separately.
4. If the particles are rocks and have some porosity, it is possible that air trapped within the particle will make it more buoyant. How two soil samples are handled and treated can cause differences in density in particles that would otherwise be considered geologically similar.
5. The measurement of any property without actually determining what causes any differences can lead to error. In soil samples, quartz and

some of the common feldspars have about the same density and appear at the same level in the column.

In light of these problems, it has been concluded that the density gradient column method, however carefully followed by the scientist, has severe limitations. The meticulous studies in 1983 by Chaperlin and Howarth show conclusively that the technique cannot identify the origin of soils on a comparative basis. Their observations are consistent with those of Frenkel in 1968, who wrote, "Published results on the pattern of a soil in a density gradient tube are sufficient to indicate, but are not sufficient to demonstrate, that this may be a significant forensic property." After considering the arguments, we can conclude that results from the method should not be the sole basis for determining comparison. It is questionable whether the procedure is even worth doing. The density gradient column method was developed in an effort to produce a pattern of particle distribution in the tubes equivalent to a "fingerprint" of soil, which could be easily and convincingly presented to a jury. The problem with the method is that, because it does not identify minerals and particles, it does not take advantage of the basis of modern forensic soil examination—the vast diversity of minerals, rocks, and fossils. But some professionals still emphasize the advantages of the density gradient column method: You can train anyone to do it, it produces pretty pictures for a jury, and nobody has to think.

8
Geophysical Instruments and the Search for Bodies

Most of the forensic geologic examples and methods so far discussed serve, in one way or another, to determine comparison. They are part of the effort to discover whether two samples have a common source. Did they once exist together? The principle underlying these studies is the usefulness of data obtained from questioned samples when compared with other data from known or control samples. And the purpose of these studies is to contribute scientific evidence that helps establish the guilt or innocence of an individual with respect to a criminal act; to help find clues during investigations; and to assist in establishing responsibility in criminal or civil matters.

There are additional tools and methods available to the forensic geologist that have been applied or might be applied to special forensic problems. In some of these examples the investigator faced with a problem that involves the earth or earth materials recognizes that geology or earth science might make a contribution and approaches an earth scientist. There are undoubtedly many possibilities other than the ones examined here, and new examples appear each year. They appear when a problem faced by the skilled, imaginative, dedicated investigator is brought to the attention of a like-minded scientist. With these types of cases, it is well to remember that the particular methods used in one case may not necessarily be useful in all similar cases. However, if the method does prove useful in another case at least the "wheel did not have to be invented again."

Magnetometer

Many geophysical instruments can give us information about the kinds of rocks under the surface of the earth, and their shape and inclinations. These have been valuable in the search for minerals, oil, and gas. Such instruments measure the properties of rocks beneath the surface of the earth and permit us to identify or predict what will be found.

The magnetometer is an instrument that measures variations in the strength of the earth's magnetic field. There are various types of magnetometers, including the magnetic balance, fluxgate magnetometer, and nuclear resonance magnetometer. Magnetic field intensity is measured in *oersteds*. The term *gauss* is also used and is numerically equivalent. The total magnetic field of the earth—that is, the field that causes a common compass to point toward the north magnetic pole—has a strength of approximately 0.5 oersteds.

The natural magnetic intensity of the earth can change several thousand gammas over distances of a few miles as the makeup of underlying rocks changes. Some rocks have greater magnetic susceptibility than others. In general, those rocks with magnetic minerals such as magnetite have greater susceptibility and thus produce more intensely magnetic readings. Some of the world's most important mines, particularly iron mines, have been discovered with a magnetometer.

A mass of iron in or on the earth intensifies the earth's natural magnetic field. If strong enough, this intensification can be detected with a magnetometer. Instruments are now available that can be carried on land, flown in airplanes, or dragged underwater by boats. The earliest use of airborne magnetometers was in submarine detection during World War II. A fluxgate type was used and formed the basis of the Magnetic Airborne Detector (MAD) system. The aircraft carrying the instrument flew over the oceans and recorded the intensity of the earth's magnetic field. When it flew over a submarine—a large mass of iron in the sea—the instrument recorded higher values. Many other large masses of iron on the bottom of the sea were also detected. Experience and careful recording of the location of known sunken ships improved the interpretation and detection of submarines. After the war the airborne magnetometer was widely used for the detection of minerals and also for locating rock structures that might contain oil.

There have been many applications of the magnetometer in forensic work. In one somewhat recent case, a well-known citizen of a Midwestern town disappeared in his new Cadillac automobile. A few months later, someone remembered seeing the vehicle near an active open-pit coal mine. There was reason to believe the missing man had been the victim of foul play. In an open-pit mine, large volumes of earth are removed from the surface, exposing the coal tens or hundreds of feet below the surface for removal. Overlying earth is carried away on giant conveyor belts and dropped in large piles away from the area to be mined. Investigators suspected that the vehicle, possibly with the victim inside, had been driven under the unattended end of the conveyor belt and buried. Company records indicated the general area of dumping during the time of the disappearance. By now, however, the area was covered by several acres of earth many feet thick. A magnetometer survey was planned. Prior to the survey, an automobile of similar kind was placed in the mine near the edge and measurements were made on the surface above the vehicle. It was determined that the instrument could detect the car even if buried up to 70 feet. When it was run, the survey produced several areas of high magnetic intensity. While none seemed high enough to indicate a mass of iron as large as a car, they were drilled. In this case, the lead proved false. The magnetometer uncovered iron wire only. The Cadillac was not found.

Metal Detector

Although they operate on several different principles, common metal or mineral detectors are used to locate small metal objects, such as firearms, that have been buried just below the surface. These instruments generally must be close to the object in order to detect it and are most useful in locating objects that lie within a few inches or feet of the earth's surface.

Seismograph

The seismograph records vibrations that travel as waves through the rocks of the earth. It is used to determine the kinds and distribution of rocks under the surface of the earth. To use the instrument, there must be (1) some way of producing shock waves and (2) a way to detect the waves after they have traveled through the earth. Because waves travel through different kinds of rocks at different velocities, by measuring the time it takes for waves to travel the known distance from the place of the shock to the place of the detector we can determine the velocity and thus something about the rock. The shock may be produced naturally by an earthquake, or artificially, for example by explosives. Small instruments use a shock produced by hitting a metal plate placed on the ground with a sledgehammer. In another method, after the shock-producing explosion at the surface, the waves travel down into the earth, reflect off rock layers, and return to the surface. The investigator records the total travel time down and back. Knowing the kinds of rocks and the velocity at which waves travel through them, it is possible to determine the depth below the surface at which the layer of rock that reflected the waves exists. In this way it is possible to find the depth below the surface at which various rocks will be found.

In some cases investigators have tried to determine the exact time of occurrence by consulting the records of seismographs that normally record earthquakes. Success depends on both the intensity of the blast and the nearness of the recording seismograph. Similar methods are used to record, on a global scale, nuclear explosions.

Keith Koper of the University of Arizona reports that information obtained from seismic data can help characterize a bomb. On the morning of August 7, 1998, terrorists exploded truck bombs nearly simultaneously at American embassies in Nairobi, Kenya, and Dar es Salaam, Tanzania. The attacks killed at least 220 people and wounded more than 4,000. In both cases, the terrorists were unable to penetrate embassy security measures completely. Still, the blasts were large enough to cause irreparable damage to two structures and completely destroy several others, along with nearby vehicles. Though the U.S. embassy in Nairobi was left standing, the nearby Ufundi building was totally demolished.

A three-component, broadband seismometer operated by the geology department of the University of Nairobi recorded the Nairobi attack. Located almost 2 miles northwest of the blast site, the instrument provided a high-quality record of the explosion, showing clear P and Rayleigh waves, as well as a series of air blasts. Scientists were able to pinpoint the precise time the explosion originated and the amount of blast energy partitioned into seismic ground motion. Furthermore, calculation of the travel time of the primary air blast constrains the total yield of the explosion since the air blast travels as a shock wave. This information helped investigators characterize the bomb.

The detector that picks up the waves is often called a geophone. Geophones form the basis of a seismic intrusion detection system. In this system, very sensitive geophones are placed in a perimeter around an area. Small boxes with a spike pushed into the ground, they are usually connected by wire or transmitter to a recorder. Current instruments are sensitive enough to record the presence of a person of average weight walking 30 to 50 feet from the geophone. Soil type and background noise from nearby roads and railroads affect the actual sensitivity of the system.

Ground Penetrating Radar

The search for human bodies and other buried objects is a major part of forensic geology. The application of geophysical methods in the search for buried bodies has had considerable success. Ground penetrating radar is a common technique. This system includes a radio transmitter and receiver connected to a pair of antennas attached to the ground. The transmitted signal penetrates a short distance into the ground. Some of the signal reflects off any object with different electrical properties than the host soil. That part of the radio signal arrives a little later in time. Fish finders work similarly in water. The instrument plots the echo from an object as an image on a screen. As the radio transmitter is moved along the ground, each new echo is plotted alongside previous ones. The result is a pattern that can be interpreted. In the Denver area, G. C. Davenport and others have researched this technique in a number of ways, including projects in which pig carcasses were buried. Investigators followed up from time to time with an instrumental search to see what patterns could be detected as the bodies decomposed. Studies by the "E" Division of the Royal Canadian Mounted Police and Professor Mark Skinner of Simon Fraser University in British Columbia were also successful. In the British Columbia studies, two goats and a bear were buried in a certain area for five years. Study participants first reduced the size of the search area by looking for disturbance and young vegetation. At that point, ground penetrating radar easily located the three burials.

Laurance Donnelly developed the following chart for a book being published by Wiley. This chart shows attributes of victim homicide disposal in shallow graves. Donnelly has long advocated the use of all geologic information in searches and avoidance of disturbing potential sites with masses of untrained people. In addition he has developed a system of maps, called diggability maps, that guide police in searches. Assuming that the person who created a shallow grave for a body or other object was in a hurry and picked a place where it was easy to dig, then the police should have maps that show where it is easy to dig. This idea follows on maps produced during World War I by the British army that showed where in France it was easy to dig trenches.

In addition, Donnelly has suggested the location where an offender chooses to dig a burial or hide may, consciously or subconsciously, be influenced by several factors such as the following:

- Type and thickness of soil/groundwater (i.e. the geology)
- Concealment from view and low witness potential
- Time of day or year· Weather conditions at the time of burial
- A place which may be known to the offender (this will facilitate navigation of escape routes, aid grave/burial location by recognition of physical features, and potentially provide an explanation if compromised during or after the act of burial)
- The ease of access on foot or by vehicle to the burial/hide location
- Diggability and the principle of least effort

Ground penetrating radar (GPR) surveying at Mount Vernon, Virginia —COURTESY OF MARGARET WATTERS, HP VISUAL AND SPATIAL TECHNOLOGY CENTRE, INSTITUTE OF ARCHAEOLOGY AND ANTIQUITY, UNIVERSITY OF BIRMINGHAM, UK

ATTRIBUTE	CHARACTERISTICS
Diggability & principles of least effort	Minimizing the time spent digging and reinstating a grave may be significant for the offender. The offender may choose a location that offers the opportunity to quickly dig an excavation of length, width, and depth for the victim to be concealed. For example, softer soils such as peat and sand may offer better alternatives than clay- and gravel-rich soils.
Bulking, excess soils, and scarring	Digging shallow graves may result in an excess of soil caused by bulking (the expansion of the soil following its removal and the volume taken up by the deceased. Digging and excess soil may produce scarring on the ground but this may become less apparent with time due to weathering or vegetation growth.
Concealment	The location of the grave may have a low witness potential and is likely to be out of view from the public or passers-by to enable the covert burial to take place.
Familiar locations	"We go where we know" and therefore the location of a grave may be familiar to the offender. If the offender is seen or compromised he may have an explanation for his presence at that particular location. Familiar locations will also facilitate navigation (particularly if at night) and escape routes. This will usually be aided by the recognition of physical features of the landscape.

In 1997 the United Nations International Criminal Tribunal for the former Yugoslavia began exhumation of mass grave sites located in northeastern Bosnia. The graves were associated with a massacre of civilians in and around Srebrenica that occurred in July 1995. However, intelligence indicated that the graves had been exhumed and the bodies moved to a number of secondary graves to disguise the massacre. Samples of soil were collected at the five primary graves and the nineteen secondary graves. The soils were sufficiently different that it was possible to demonstrate the body transfer and this evidence was submitted in what was to be a successful prosecution.

Global Positioning System (GPS)

The global positioning system (GPS) is a global navigation system composed of a network of communication satellites (thirty-one as of 2011) placed into medium earth orbit for the U.S. Department of Defense and monitored by a series of ground-based tracking stations operated by the U.S. Air Force and the National Geospatial-Intelligence Agency. Instruments that provide location and route data are available in many stores and are in common use. There are many cases where these instruments have been attached to a suspect vehicle and the information identifying where the vehicle had been and the route it took was useful in solving a crime.

Geiger Counter

Many minerals are naturally radioactive. Among these are ores of uranium and thorium, and many geologists are employed to explore for these ores. Fortunately, two common instruments, the Geiger counter and the scintillation counter, detect radioactivity directly. In forensic work, radioactivity comes into play in several ways. Sometimes investigators must detect radioactive minerals. In addition, it is possible to use radioactive powders and pastes, detectable later with a counter, to show that a person or object was in contact with the substance at some point.

In a classic case in the late 1930s, lead bars were stolen from the Palmer Physical Laboratory in Princeton, New Jersey. These bars were not ordinary lead bars; they contained radioactive cobalt. Assuming the thief disposed of the lead by selling it to a junkyard, investigators searched all such places in the area with a Geiger counter and recovered the lead.

Forensic Applications of Geophysical Techniques

	Metal Detector	Magnetics	Electro-Magnetics	Ground Penetrating Radar	Comments
Shell casings	●				
Bullets	●				
Unexploded ordnance	○	●			Two-box metal detector
Grave (shallow)		○	◍	●	
Grave (under concrete)				●	Depends on reinforcing mesh
Body underwater				○	Consider fathometer
Property in water	◍	◍		○	GPR in freshwater only
55-gallon drum 5–10 feet deep		●	◍	○	

● Most Applicable ◍ Applicable ○ May Work

Usefulness of geophysical techniques to the forensic geologist —COURTESY OF G. C. DAVENPORT

Carbon-14 Dating

Radiocarbon dating or carbon-14 dating was developed just after World War II. This method relies on the decay of carbon-14 in the tissues of a plant or an animal that has died. It can produce dates back to 70,000 years. Archeologists and paleontologists use the method to date sites and remains of early humans and animals. Nuclear weapons testing changed the amount of radiocarbon present, so a correction must be made when relatively recent dates are attempted. However, this permits accurate measurements of material that is not too old. In many cases, radiocarbon dating has helped determine the age of human bones. Usually this is for establishing whether the bones are young enough to belong to victims of prosecutable crimes. The technique has also been used to determine the actual age of wooden furniture, cloth, and painted canvases in cases of suspected fraud.

Scientists have long searched for a radiogenic isotope with a short half-life that becomes incorporated in the body. If such a material could be found and documented, it could be used to determine time since death. Stuart Black of the University of Reading in the United Kingdom is exploring a method involving the daughter isotopes of lead and polonium-210. These form in the atmosphere and are incorporated in food and water. Humans bring them into their bodies while breathing, eating, and drinking. Studies of the method have involved measurements of elephant bone of known age in Kenya.

Fluorescence

In ultraviolet light, many minerals glow with a color different from their color in normal white light. Some fluorescent materials are almost invisible in normal white light. Ultraviolet light has long been used to identify minerals that fluoresce. Many museums have exhibits of spectacular fluorescence.

Like all electromagnetic radiation, ultraviolet light travels in wave form. Electromagnetic wavelengths vary over a large spectrum, from the longest radio waves of 12.5 miles to cosmic rays shorter than 0.0004 angstroms in length. The light we see ranges from violet, at approximately 3,800 angstroms, to red, at approximately 8,000 angstroms. The ultraviolet range of wavelengths lies between approximately 1,800 and 4,000 angstroms. Commercially available ultraviolet lamps produce either short radiation from 1,800- to 3,000-angstrom wavelengths or long radiation from 3,000- to 4,000-angstrom wavelengths. Both these ranges are invisible to the human eye; the light that is commonly seen from these lamps is violet light that has not been removed. This visible light is useful because it tells us the lamp is on and working.

When a fluorescent material is exposed to ultraviolet light, the atoms in the material become excited, and the electrons traveling in small orbits around the nucleus of each atom jump to another more distant orbit. Electrons in the outer orbit drop to replace electrons that moved. The movement of these electrons produces energy, and the result is the generation of light visible to the human eye. Thus materials that fluoresce glow with visible light when excited with invisible light and may be seen and photographed. The fluorescence, which lasts only as long as the material is exposed to ultraviolet light, may appear as shades of blue, brown, green, orange, gold, red, white, yellow, violet, or purple. The color depends on the material and in most cases is quite diagnostic. For example, rock chips that come to the surface during the drilling of an oil well are examined under ultraviolet light because most oils fluoresce with distinctive colors. The oil sticks to the rock chips and gives clues as to the kind of oil below.

In addition, it is possible to obtain identification kits with pastes, powders, and sprays that fluoresce in a variety of greens, blues, and yellows. The fluorescence on a suspect who has touched objects sprayed or coated with one of these materials is easy to spot; the investigator need only be familiar with the color used. If there is doubt, comparison with known samples of the fluorescent material can be made. While both long- and shortwave fluorescent light excites most of these materials, the type of light appropriate for a material should be checked before use. Long-wave light is common for many of the commercially available materials because short-wave ultraviolet light can seriously damage the eyes and must be used with care.

Fluorescent paste is commonly placed on tripping levers of fire alarm boxes as a means of identifying people who set off false alarms. In one case, a suspect was apprehended while running away from the scene of a false fire alarm. His fingers were examined under ultraviolet light and the skin showed a strong fluorescence, which was stated at his trial. However, a second examination two days after the suspect's apprehension revealed that he wore a sport coat made of synthetic fibers with fluorescent dyes. The loose fibers and lint from the coat, stuck to his hands by perspiration, were what fluoresced. This was confirmed by microscope examination of fibers from the suspect's hand. The first identification was shown to be in error and the suspect's innocence was established. To be certain the suspect had not had paste on his hands and deliberately or accidentally removed it in the time between examinations, a sample of the paste was tested. The material was still fluorescent and easily identifiable on skin after several days.

On occasion, finely ground, distinctive minerals are mixed with a fluorescent paste or powder for use at a single location, resulting in a truly individual identification. If a person comes into contact with that specific paste in that specific place, there is no doubt about it.

Some rubber products such as tires are manufactured with oils that fluoresce under ultraviolet light. In some cases it has been possible, with ultraviolet light, to recognize the imprint of an automobile tire on a clean concrete floor. Oil from the tire remains on the floor in much the same way oils from a human finger remain on a surface in a fingerprint. There has been some success in recording the intensity of ultraviolet light in the different ultraviolet wavelengths, thus characterizing the oils in certain tires. This offers the possibility of identifying the rubber products of different manufacturers from the impressions.

Water Currents

In general, in oceans, bays, and lakes, water moves along the bottom as a current. The direction and velocity of these currents may or may not be the same as the direction and velocity of floating objects, moved for the most part by wind. In fact, at different depths in the water mass, currents commonly have different velocities and directions. Geologists and oceanographers have long measured the velocity and direction of currents in oceans, bays, and lakes. The geologic purpose of these measurements is to predict the movement of sediment, sand, and mud.

Scientists use various methods to measure current direction and velocity. Drifter surveys measure the movement of currents along the bottom of water bodies. Drifters are disks, usually made of plastic weighted just enough to rest on the bottom and move with the current. They are placed in the water at a known time and later recovered. Figuring the time it took for the disks to travel to the place of recovery reveals how fast they drifted and thus the velocity of the current that moved them. Similar measurements can be made by placing dye in the water and observing how fast the mass of water with the dye moves. Meters of various types also directly record the velocity of water currents. The velocity of a floating object can be measured directly by recording the time it takes to move from one place to another in a given wind.

In general, an unweighted human body is denser than water for approximately six days after death and acts as a bottom drifter. After about six days an unanchored body floats, becoming a surface drifter. In some cases, knowledge of the distribution and velocities of currents has made it possible to locate where a body entered the water. Alternatively, if time and point of entry are known, the probable whereabouts of a body or other object can sometimes be predicted.

The distribution and velocity of currents in San Francisco Bay have been studied extensively. Bottom-drifter surveys have been made. Gangland-style slayings in the San Francisco area have been followed by the disposal of bodies in the water of the bay near the Golden Gate. Bodies have been recovered along the shore at the southern end of the bay. It has been possible to estimate the time a body was dumped into the bay and the location of dumping, or to anticipate the point where a body will surface. Knowing the surface wind-driven current and bottom current directions and velocities during a given time, it is possible to predict the place on land where the body will arrive. Such studies are not restricted to human bodies and with the appropriate information may be applied to any floating or bottom-drifting object. Recent studies of bodies recovered in the tidal estuaries of Great Britain, such as the Humber and the Thames, using information on time, water depth, flow vectors, and flow paths, have been very successful in determining the time of a crime.

Aerial Photography

Aerial photographs are now used for a wide variety of purposes including mapping, surveying, highway planning, geologic and forestry studies, archeology, assessing crop damage, military operations, and stockpile inventories. The American Society of Photogrammetry lists more than one hundred ways in which aerial photographs can be used. Many archaeological discoveries have been made solely through the use of aerial photographs. In the British Isles, archeologic finds initially detected on aerial photographs date back to the Roman occupation. Aerial photographs have been used to locate modern burial sites. The use of photography to locate burial grounds or other disturbed sites rests on the principle that disturbed sites have altered soil conditions and are colonized by plants different from those growing on adjacent undisturbed areas. The soil in the disturbed area looks different from that in undisturbed areas around it. Disturbance of an area alters its moisture regime, organic matter distribution, soil structure, and possibly other properties.

During the course of an investigation, it is sometimes critical to establish the time of a certain activity—when a wetlands was filled in, a borrow pit dug, a forest cut, or a structure built or demolished. An aerial or ground photograph, with date, gives indisputable evidence of physical features or landscape conditions at a specific time. By comparing photographs of a landscape taken at different times, we may draw some conclusions about when an event took place. Aerial photographs can also be used, for example, to track the movement of material leaching from landfills. Fortunately, for establishment of baselines, such federal agencies as the U.S. Department of Agriculture, the U.S. Geologic Survey, and the military, as well as some commercial organizations, take aerial photographs of parcels of land from time to time. Virtually all modern aerial photographs record the date on the film, increasing the usefulness of the work.

Remote Sensing Data

Remote sensing data can be useful in searching for hidden cabins, illegal mining and logging, cultivation of marijuana, and other hidden objects or activities. A major source of this type of information and of maps is the Earth Resources Observation Systems (EROS) Data Center in Sioux Falls, South Dakota. The following 2001 press release from the U.S. Geological Survey illustrates the value of this type of information.

HIGH-TECH USGS MAPS BEING USED TO SOLVE XIANA MYSTERY

> Aerial photographs supplied to the Santa Clara County Sheriff's Search and Rescue Team, by the U.S. Geological Survey, are being used to search for the remains of Xiana Fairchild. A child's skull, which was found near Lexington Reservoir on January 19, has been identified through dental records and DNA tests, as that of the missing 7-year-old Vallejo girl who disappeared in December 1999.

On January 26, 2001, an official with the Western Disaster Center at Moffett Field contacted the USGS in Menlo Park to request current, detailed geographic data that could be used to assist in the search for the girl's body. The Menlo Park cartographers immediately contacted their colleagues at the Earth Resources Observation Systems (EROS) Data Center in Sioux Falls, S.D. Within hours technicians there e-mailed updated, state-of-the-art, digital photo maps of the search area directly to a Geographic Information System (GIS) analyst working with the sheriff's department.

The familiar USGS topographic maps are routinely used in search and rescue missions all over the nation. In addition, searchers are now using electronic map programs that use USGS digital orthophoto quads (DOQs), digitized aerial photographs that cover one-fourth of a topo map. They can be sent electronically to searchers who can download them on site with battery-powered laptop computers.

9 Mine, Mineral, Gem, and Art Fraud

Fraud is theft by lying. There are people who lie for profit. They lie or misrepresent many types of objects or situations, but in the areas of mining property, gems, minerals, and art, the tools of forensic geology have been very useful in identifying and assisting in prosecuting the crimes.

Alastair Ruffell reminds us the faking of objects using geologic materials is likely to have been occurring before written documentation. During Mesopotamian time (4,000 BC) fake stones were created by heating silt to a partial melt and then cooling it. Egyptian fakery using geologic materials was well established by 300 BC: a big toe made of linen, glue, and plaster (calcium sulphate hemihydrate) was dated at between 1295 and 664 BC, the earliest recorded fake body part.

The U.S. Securities and Exchange Commission's antifraud rule 10b-5, cited in most of its fraud cases, describes fraudulent activity as follows:

> Rule 10b-5: It shall be unlawful for any person, directly or indirectly, by the use of any means or instrumentality of interstate commerce, or of the mails or of any facility of any national securities exchange,
>
> - To employ any device, scheme, or artifice to defraud,
>
> - To make any untrue statement of a material fact or to omit to state a material fact necessary in order to make the statements made, in the light of the circumstances under which they were made, not misleading, or
>
> - To engage in any act, practice, or course of business which operates or would operate as a fraud or deceit upon any person, in connection with the purchase or sale of any security.
>
> (17 CFR 240.10b-5)

Note that untrue statements are equal to omitting material facts. No one knows this better than Denver forensic geologist David Abbott. Abbott worked as a forensic geologist for the Securities and Exchange Commission for many years, examining every variety of mining and mineral scheme imaginable. He says that, after several years, he could spot a potential fraud just by reading a prospectus. The same scams kept reappearing, often with few if any changes.

A common scam is the offer to own your own small gold mine. The offer is for an interest in or an entire pile of mine debris—often tailings from an abandoned mine. The pitch guarantees that there is so much gold in the pile. This is either an outright lie or is backed by the analysis of a dishonest assayer. In the old days of the mine's operation, according to the story, they just didn't get all the gold and threw some away into the tailings piles. Of course the value of the promised amount of gold is greater than the asking price. Buyers are given the option to visit the mine and process their gold themselves or to hire someone—a friend of the seller—to process it for them. Most folks take the second option and are pleased to find out that the processor has recovered even more gold than originally promised. Buyers then have another choice: take their gold now and pay tax or store it for a small annual fee in the seller's safe. Most choose the latter. When the day comes to collect, the buyer finds out that the seller is no longer in business and there is no gold and no safe. Variations on this scam are endless. The forensic geologist examining such cases usually goes to

Piles of mine tailings sold in a scam. The sticks record the buyers' and owners' names.
—COURTESY OF DAVID ABBOTT

the site and analyzes the dirt piles. The fraud usually lies in the statements about the value of gold in the pile.

Abbott tells of another scam he has investigated that also has many variations and reincarnations. There is a black-sand beach on the west coast of Costa Rica that purportedly contains a concentration of gold. The black sands are trucked to San Jose in the central highlands for processing. The prospective investor is shown the processing plant and the trucks being loaded at the beach. This scam includes a false statement about the amount of gold in a truck of sand. Abbott demonstrated this by panning samples from the beach and determining the actual amount of gold present. In addition the promoters point out that the beach sand is constantly replenished, so the money will roll in forever. They fail to say the rate at which the beach sand is replenished with heavy minerals and whether that rate is sufficient to keep up with the rate of removal.

Lots of mine scams result from salting a property with gold. The metal may be sprinkled around or shot from a shotgun. Examinations of such cases seek to answer the question: Is this the way gold from this property should look, or is this gold brought in from somewhere else? Under the microscope the salted gold is often revealed to be gold filings with file marks or chopped-up gold foil.

The most massive and disastrous mine fraud in recent times is Bre-X. The Bre-X scandal began with a near-worthless gold mining property along Busang Creek on the island of Borneo and grew to become the biggest mining fraud in Canadian history. John Felderhof and Michael de Guzman were geologists. Both had excellent professional reputations. In 1992, both were desperate for work. Felderhof sent de Guzman to Busang to see if it could be sold. The resulting report was very optimistic. Canadian mining entrepreneur David Walsh had formed a company in 1989 which he called Bre-X for his oldest son Brett. It was listed on the Alberta Stock Exchange and traded at between fourteen and thirty cents for four years. The company was not doing very well when, in 1993, it bought Busang for $80,000 on the basis of de Guzman's report. Drilling began a few months later, producing little gold from the first two holes. The third hole produced significant gold as did most of the later holes. Bre-X stock went to $270.

Ownership of the property was more complex than the geology. Ultimately, in 1997 American mining corporation Freeport-McMoRan acquired a 15 percent share. Through a Suharto holding company, the government of Indonesia held 40 percent, and Bre-X kept 40 percent. Bre-X stock was now traded on the Toronto Stock Exchange and the NASDAQ.

Finally Freeport and others took a careful look at the deposit. It turned out the third drill hole had been salted with a man-made gold-and-copper alloy. The other drill samples contained placer gold purchased from local miners. The core samples had all been thrown away, and a fire in one of the buildings had destroyed all records. On the night of March 10, 1997, on which David Walsh and John Felderhof were to be honored as "Canadian Prospectors of the Year" by the Prospectors Association, the two men received calls from Freeport-McMoRan telling them that the drilling had shown insignificant amounts of gold. Freeport wanted an explanation. Nine days later, in the Bornean jungle, Michael de Guzman fell from a helicopter under mysterious circumstances. His death was reported as a suicide. Billions of dollars had been lost, and Busang was once again a worthless mining property. It is not known if Felderhof knew of the salting scam, but he made a substantial profit selling his stock and now lives in the Cayman Islands. Investigators believe that Walsh may not have known of the scam. He died in 1998 of a massive heart attack. We may never know the full details of the biggest mining fraud in Canadian history.

Gem Fraud and Misrepresentation

For centuries people have sold or traded gems and represented them as something other than what they are. Sometimes this results from their lack of knowledge, lack of instrumentation for accurate identification, or reliance on others. In many cases, it is simply lying for profit. In the diamond market, one kind of fraud is claiming that a diamond is natural rather than synthetic, or "clarity-enhanced." Clarity enhancement is filling fissures in a diamond with a solid substance, such as glass, that reacts to light in ways similar to diamond. As a result, the fissures become less visible, improving the apparent clarity of the diamond. In most cases of diamond fraud, another synthetic material is sold as diamond. One such material is YAG (yttrium aluminium garnet), marketed under such names as Diamonaire. Another is strontium titanate, sold as Fabulite or Diagem. The best diamond imitation, which became available in 1977, is Zirconia, also called fianite, phianite, and KSZ. It is really an yttrium zirconium oxide.

The ancient Egyptians were known to substitute glass because the real gems were rare and expensive. In 1758, Joseph Strasser in Vienna developed a high-refractive-index glass very similar to diamond in appearance. The Empress Maria Theresa outlawed its production and sale, but it reached the European market through Paris. The ability to produce artificial stones that are chemically and physically the same as the natural material was achieved in 1888 by the French chemist A. V. Verneuil, who produced rubies at a commercial price. Earlier attempts succeeded in producing synthetic stones, but they were only scientific curiosities. It has been argued that synthetic emerald is far more pleasing to the eye than almost all of the natural stones, and that a natural Burmese ruby cannot be distinguished from a Rock Creek, Montana, ruby despite the much higher price of the Burmese. The reality is, people buy gems for many reasons: for marketing, investment value, snob appeal, fashion, as birthstones, and in some cases, just because they like them. A good example of marketing is the popularity of black diamonds. These stones, so filled with inclusions that a few years ago they sold only as abrasives, have now become treasured objects of finery. Crime occurs when someone, for profit, represents a stone as something it is not. There is more misrepresentation in the colored-gem market than in the diamond market, primarily because colored gemstones are more complex.

With modern equipment and knowledge of crystals, it is now possible to "create" or "grow" almost any gemstone. Synthetic amethyst, alexandrite, ruby, emerald, sapphire, opal, and even turquoise are on the market. Many synthetics are themselves expensive, and most have become more difficult to distinguish from their natural counterparts, resulting in inadvertent misrepresentation. Some, quite beautiful, are marketed as synthetic.

Simulated or imitation stones should not be confused with synthetics, which possess essentially the same physical, chemical, and optical properties as the natural gem. A simulated stone is usually a very inexpensive imitation that resembles the natural stone in color but little else. Many imitations are glass, but they can also be plastic. Imitations, or simulants as they are also called, are very easily differentiated from the genuine by careful visual examination and simple gemological testing. Glass simulations exist of all the colored stones, and glass and plastic simulated pearls, turquoise, and amber are also common.

Another form of deception is the misrepresentation of a more common, less expensive stone as a rarer, more expensive gem of similar color. Garnets may be presented as rubies, diopside as peridot, or green-dyed chalcedony as malachite. Today, as more and more natural gemstones in a wide variety of colors enter the market, deliberate and accidental misrepresentations can occur.

Color enhancement of gemstones is not new. Many techniques, used for generations, do not in themselves constitute fraud. Because color enhancement is so common, it's important to understand exactly which procedures are acceptable in the industry and which represent deceptive or fraudulent practices aimed at passing off inferior stones as more expensive ones.

Subjecting stones to sophisticated heating procedures is the most common method of changing or enhancing a gem's color. Heat treatment is used routinely on a variety of gems to lighten, darken, or completely change color. It is generally not fraudulent when used routinely on certain gems and when the results are permanent. This procedure is an accepted practice routinely applied to the following stones:

Amber: To deepen color and add "sun spangles"

Amethyst: To lighten color; to change the color of pale material to "yellow" stones sold as citrine

Aquamarine: To deepen color and remove any greenish undertone for a "bluer" blue

Carnelian: To produce color

Citrine: Often produced by heating other varieties of quartz

Kunzite: To improve color

Morganite: To change color from orange to pinkish

Sapphire: To lighten or intensify color; to improve uniformity

Tanzanite: To produce a more desirable shade of blue

Topaz: In combination with radiation, to produce shades of blue; to produce pink

Tourmaline: To lighten darker shades, usually of green and blue varieties

Zircon: To produce red, blue, or colorless stones

The color obtained by these heating procedures is usually permanent. However, heat treating, which can be detected with a microscope, is best disclosed unless it is widely known that it is a standard practice in the trade. Sometimes heat treating is done in a chemical atmosphere that allows atoms of a chemical to diffuse into the outer part of the stone. This really does change the composition of the outer material and should be disclosed. The changed outer layer can usually be polished off, exposing the underlying, unchanged stone.

Radiation techniques produced by several methods are now in common use. Each method has a specific application. Sometimes radiation is used in combination with heat treatment. As long as the technique produces stable results, color enhancement by radiation techniques is not considered fraudulent. However, the U.S. Federal Trade Commission believes sellers should disclose the fact that a stone has been treated. Radiation techniques are routinely used for the following stones and do not constitute fraud:

Diamond: To change the color from off-white color to a fancy color (e.g., green, yellow)

Kunzite: To darken color

Pearl: To produce blue and shades of gray ("black" pearls)

Topaz: To change from colorless to blue; to intensify yellow and orange shades and create green

Tourmaline: To intensify pink, red, and purple shades

Yellow beryl: To create yellow color

Painting is often used with cabochon transparent or semitransparent opals to create a stone that looks like precious black opal. A black cement or paint is spread on the inside of the setting. When the opal is placed inside, light entering it gets trapped and reflected back, giving the opal the appearance of a fine black opal.

Foil-backed stones are not frequently seen in modern jewelry but are relatively common in antique jewelry. Foil backing is used with both non-faceted and faceted stones, usually set in a closed-back mounting. The inside of the setting is lined either with silver or gold foil to add brilliance and sparkle, or with colored foil to change or enhance color.

Smoking is a technique used only with opals. It gives off-white to tan opals from Mexico a more desirable, moderately dark coffee-brown color that greatly enhances the opal fire. A cut, polished opal is wrapped tightly in brown paper and placed in a covered container over moderate heat until the paper is completely charred. When cooled and removed, the opal has a much more intense brown body-color and fire. But if this smoke-produced color coating is badly scratched, the underlying color shows through. The stone has to be resmoked. This treatment is easily detected by wetting the stone, as with saliva. When wet, some of the fire disappears, but reappears when the surface dries.

Fractures in the surface of a colored gem can be filled with a liquid glass or glasslike substance, or with an epoxy resin-type filler. Fillers make cracks less visible and improve a stone's overall appearance. A coloring agent added to the filler can simultaneously improve a stone's color. Selling a filled gem without disclosure is not an accepted trade practice. To do so knowingly constitutes fraud. Nonetheless, the number of glass-filled rubies encountered in the marketplace and sold without disclosure has increased dramatically. Emeralds filled with epoxy resin are now in wide circulation and often sold without disclosure. Some dealers still routinely oil emeralds.

Composite stones are composed of more than one part. There are two basic types. Doublets are composite stones that have two parts, sometimes held together by a colored bonding agent. Triplets consist of three parts, usually glued together to a colored middle part.

Doublets were widely used in antique jewelry before synthetics were developed, and are still fairly common. In antique pieces, the most commonly encountered doublet consists of a red garnet top fused to an appropriately colored glass bottom. Another type of doublet is made from two parts of a colorless material, fused together with a colored bonding agent. An "emerald" (sometimes sold as a "soudé emerald") can be made from a top and bottom of colorless synthetic spinel that are held together in the middle by green glue. Red glue or blue glue simulates ruby or sapphire.

Triplet composites are frequent in the opal market and have substantially replaced the doublet there. The triplet is exactly like the opal doublet except that a cabochon-shaped colorless quartz cap covers the entire doublet, protecting the delicate doublet from breakage and giving the stone greater luminescence. Some of these are disclosed and marketed at a higher price as improvements over natural opals.

Some of the above observations on enhancement and fraud come from the Gem Shop of Delhi, India.

Amber has been the subject of many cases of fraud. Most amber started out 30 to 90 million years ago as sap seeping from trees. The sticky, aromatic resin oozed down the trees' sides, attracting and trapping seeds, leaves, feathers, and insects. With burial and time, the resin hardened through natural polymerization into the gold-colored gem called amber. Amber occurs all over the world, but the Baltic region is perhaps best known for it. A specimen's value often depends on the fossils it contains. Small vertebrate inclusions in amber can command very high sums. Needless to say, this opens up opportunities for the amber forger. There are many methods for producing fake amber, including using copal or a synthetic resin or plastic such as unsaturated polyester resin. Another method is to dig out a small cavity in real amber, insert an animal carcass, and refill the cavity with resin. The completeness and perfection of these creatures compared to the broken, bent ones found in most real amber can be a clue. Creatures caught in sticky sap on the side of a tree do not usually just lie down and die. About amber forgery, noted geologist and forensic geology teacher

Dr. Jack Crelling often says, "If it looks like someone laid out a perfect organism in the resin, someone probably did."

In 1900, J. P. Morgan purchased the famous Bement collection of amber for $100,000 and presented it to the American Museum of Natural History. Specimen AMNH 13704 was labeled and noted in the catalogue as "small tree toad in amber." Even when reexamined by experts in 1993, the specimen appeared to be 40-million-year-old Baltic amber. The frog was complete with the middle of the head and right eye somewhat collapsed. The skin showed some pigmentation and bones could be seen. Air bubbles, common in most forgeries between the specimen and the resin, were absent. Experts originally judged a thin crack across one end of the specimen to be natural. A laboratory study of the Bement frog with a stereo binocular microscope and fiber optic lighting revealed a very small sea scallop shell adjacent to the frog. The forger either made a mistake or had a sense of humor when he or she introduced the seashell, which apparently climbed up the tree into the sap. It turned out that someone drilled a hole in the amber, inserted the frog and seashell, then carefully cemented the removed side back along a natural fracture.

Laboratories such as the Swiss labs SSEF in Basel and Gübelin in Lucerne research the geographic sources of gems and issue country-of-origin reports. Country of origin has become more important in recent years for several reasons. Gemstones can be used to fund terrorism. Being from a particular location may increase a gem's price. Information about origin can also help in the monitoring and enforcement of world trade agreements.

In cases of stolen gem material, it may be necessary to trace stones back to a particular mine. Most of this work relies on trace elements in the stone and such internal features as inclusions in addition to physical and optical properties. Dr. Pornsawat Wathanakul of Kasetsart University in Bangkok and the Gem and Jewelry Institute of Thailand at Chulalongkorn University is one of the leading researchers in the field of determining gem sources. A laboratory such as the Gem and Jewelry Institute is not only well furnished with standard sets of gem identification devices, it also has sophisticated gem analysis instruments such as the EDAX or EDXRF (energy dispersive X-ray fluorescence) spectrometer, the FTIR (Fourier transform infrared) spectrophotometer, the UV-VIS-NIR (ultraviolet visible near infrared) spectrophotometer, the laser Raman spectroscope, a cathodoluminoscope, and an X-radiography unit. Most important are the institute's databases and other information used to determine origin. For stolen stones, in addition to determining country of origin, the institute maps stones with cathode pictures and spectroscopic data/character. The Gemological Institute of America in Carlsbad, California, has existed since 1931. World leaders in the identification of gemstones, the institute offers outstanding courses in identification.

Art Fraud

Walter C. McCrone founded the McCrone Research Institute and McCrone Associates. He was a prominent figure in the world of identifying small particles and teaching others how to do the same. In addition, he published *The Particle Atlas,* the six-volume bible of particle identification, and created the journal *Microscope.* Many people in the world of forensic geology worked or studied with him. McCrone's most famous work included efforts to determine the authenticity of such well-known objects as the Shroud of Turin and the Vinland Map. Most forensic geologic work in the area of art involves identifying minerals in pigments or ceramics, most often with the polarizing microscope.

Art forgery is unusually common. Naturally, the forger selects a famous artist who is no longer alive and attempts to duplicate the artist's style or an existing painting. In addition the forger should use materials employed at the time the artist was working. A notorious example began with the discovery in 1985 of 1,500 works Larionov pastels and drawings. Mikhail Larionov, a modern painter who left Russia in 1915, has become very fashionable in recent years and his work commands high prices. Experts assumed Larionov left the discovered works in Russia when he moved to Paris. The collection was widely distributed and exhibited in Germany and Switzerland. Ultimately, questions arose as to whether the 1,500 works were genuine. With the polarizing microscope, McCrone examined two of the pastels and found titanium white, with the titanium in the mineral rutile form. He confirmed the identification with X-ray diffraction. McCrone knew that artists began using this material in the 1940s. In addition, an absence of barium sulfate put the date in the 1950s. The allegedly discovered Larionovs were fakes. X-ray diffraction patterns were similar for material from several paintings, making it highly likely that they were done at the same time by the same artist with the same pigment mix.

Only two completed paintings are universally accepted as the work of Leonardo da Vinci: *The Last Supper* and the *Mona Lisa.* In 1985, John Harrington purchased a painting now known as *Christ Among the Doctors.* Historical evidence—numerous similar paintings—suggested that the piece might be a da Vinci. In the art world, many similar paintings can point to an original, of which they are copies—in this case, a lost da Vinci. Walter McCrone had the opportunity to study the painting. The problem here was not to show it was a fraud but rather, through forensic examination, to provide evidence that it could be an original. Unfortunately, while certain evidence could indicate the painting was done during da Vinci's lifetime, it could not prove da Vinci himself painted it.

First, McCrone had radiocarbon dates run on the linen canvas of the painting. He got an average date of 1495, plus or minus 29 years—consistent with da Vinci's active painting period. Examination of the pigment with the polarizing microscope and other analytical instruments revealed materials in use by da Vinci and others of his time. McCrone's work was done. Further study by art experts concurred; the piece could be Leonardo's third complete painting.

10

Forensic Geology: Yesterday, Today, and Tomorrow

By now, soils and other earth materials have long been used in forensic matters. Expert testimony in this area is admissible in most jurisdictions and has made many contributions to justice. In the 1973 first edition of their classic book on scientific evidence, Moenssens, Moses, and Inbau cited the following criminal cases in which information from soil or related material was admitted as evidence.

State* v. *Baldwin 47 N.J. 379, 221 A.2d 199 (1966) *petition for certif.* to App. Div. denied, 246 A.2d *459* (1968), cert. denied 385 U.S. 980.	Soil from crime scene compared with soil from defendant's car.
State* v. *Spring *Supra* n. 40.	Soil on boots found to have evidential value.
State* v. *Atkinson 275 N.C. 288, 167 S.E.2d 241 (1969), remanded for resentencing, 183 S.E.2d 106 (1971).	Soil on shovel at defendant's home compared with soil from victim's burial scene.
Territory* v. *Young 32 Hawaii 628 (1933).	Soil on defendant's trousers compared with soil at rape scene. Defendant's alibi location produced samples that did not compare with soil on his trousers.
State v. Coolidge 109 N.H. 403, 260 A.2d 547 (1969) rev'd on other grounds 403 U.S. 443 (1971). Forty sets of particles were matched microscopically with regard to color, hue, and texture. Instrumentation found at least 27 sets to be indistinguishable in all tests.	Particles removed from victim's clothes compared to particles in suspect's automobile.

Aaron* v. *State 271 Ala. 70, 122 So.2d 360 (1960) pet. for writ error denied 275 Ala. 377,155 So.2d 334 (1963).	Dust from wallboard broken off during a rape compared with dust on the clothes of the defendant.
People* v. *Smith 142 Cal. App.2d 287, 298 P.2d 540 (1956).	Plaster dust implicated burglary defendant.
State* v. *Washington 335 S.W.2d 23 (Mo. 1960).	Mortar particles on defendant's clothes compared with mortar found in the burglary access hole.

The study of rocks, minerals, soils, and related material has many uses in forensic work. Because of the very large number of kinds and combinations of these materials, their potential as physical evidence is high. In some cases their value approaches that of individual evidence such as fingerprints. However, two factors limit the use of these materials both in investigations and as evidence:

1. They must be present in sufficient quantity for analysis.
2. The investigator must recognize their potential value and collect them for analysis.

With certain manufactured products such as narcotics or explosives, or human products such as blood or semen, chemical identification alone may have evidential value. But earth materials require far more study and professional judgment. In addition, scientists can rarely make physical matches that show positively that one sample was once part of another. This can sometimes be done with glass fragments or concrete blocks that were once part of a single piece. For example, during a strike, several rocks were thrown through the windshield of a truck, causing serious damage to property and persons. Investigators found several rock chips in the suspect's vehicle. The rocks that inflicted the damage were also available for study. They were limestone with abundant impressions of the fossils of ancient shellfish. The rock chips from the suspect's vehicle contained fossils of the same type as those in the limestone. In fact, some of the fossils in the chips fit perfectly into fossil impressions in the limestone, like the pieces of a jigsaw puzzle. This left no doubt: the two had been part of the same rock. But such absolute individualization is only rarely possible.

To demonstrate comparison or lack of comparison, the forensic geologist must find particles of minerals, rocks, and related objects that are rare. For this reason, gross methods that measure such variables as density distribution, color, chemical composition, or size distribution may contribute to establishing comparison but are seldom sufficient in themselves for such a judgment. Instead, what is generally necessary is the study and identification of individual particles and their amounts, combined with these other methods. For that reason, the generalist has an important role in forensic science, particularly in recognizing interrelations between kinds of physical evidence, but for forensic geologic work, a trained specialist is essential.

For the same reason, a single standardized test for soils and related material as a basis for comparison is a desirable goal for research, but not likely in the immediate future. However, it is possible to outline a strategy for the study of these materials including the methods most likely to lead to sound scientific judgment. In considering strategies, types of studies divide into those that aid investigations and those that help establish comparison or lack of comparison as evidence in a court of law. The distinction is not always clear, and preliminary studies to assist an investigation may ultimately serve as evidence. The quality of work should be of the same high order in all cases.

Studies to Assist an Investigation

The forensic geologist often receives samples associated with crimes or suspects for study. The scientist studies the material in the hope of identifying rocks, minerals, or other particles that will help in an investigation. The key question is, where did this material come from? After studying the material and identifying the particles, the geologist can usually outline possible sources—for example, safe insulation or a certain geographic location. The solidity of the answer depends on how unusual the material is and on the ability and experience of the scientist. In this type of study, the stereo binocular microscope is normally the first instrument used. Examination with this instrument provides information about the types of material in samples. Next, the scientist will probably use other methods: the scanning electron microscope, thin sections under petrographic microscope, or X-ray diffraction.

Knowledge of the local geology and mineral industry can be essential in studies that aid investigations. In several cases in which they helped law enforcement, geologists in the California Division of Mines and Geology under the direction of then state geologist James F. Davis have proven this true. In a kidnapping case, an assemblage of marine and freshwater diatoms on the floor of a getaway car suggested a diatomite processing facility or storage area, narrowing the field of suspects for the police. The kidnappers were eventually caught and convicted. In another case involving larceny of a Cadillac, the thieves ran out of gas and abandoned the car. When apprehended, they did not know the location of the abandoned car. However, they reported hitching a ride with a miner who held chromite, mercury, and bentonite claims. Since these materials occur together in only one known area in California, geologists led investigators to the Cadillac.

In another example, Gerald Frank Stanley, convicted of murdering his second wife in 1975, was imprisoned in California. Released after four years, he remarried. In 1980, he was convicted of murdering his fourth wife. The day before Stanley's fourth wife was murdered, a young woman accepted a ride from Stanley. Her body was later found at an oil-well site in northern California. The prosecution used this crime to argue that Stanley should receive the death penalty for the murder of his fourth wife. Evidence tying Stanley to the victim at the oil-well site included the type of "geologic accident" that commonly appears in forensic geology. The apron of crushed rock and gravel around the oil well contained local gravel. Similar material lay on the floorboard of Stanley's automobile. However, extraneous, nonlocal rock fragments were also present in the apron. Police learned that a load of crushed aggregate hauled from over 300 miles south had also been spread at the well site. These rock particles also compared with particles in the vehicle. Although Stanley was never tried for or convicted of the oil-well site murder, this information contributed to the jury's decision to recommend the death penalty for the killing of his fourth wife.

Studies to Establish Comparison or Lack of Comparison

The purpose of comparison studies is to establish with a very high degree of probability that a given sample is similar to or dissimilar from another sample and that, with an equally high degree of probability, the two samples came from, or could not have come from, the same small area. In challenging this type of evidence two questions are normally asked:

1. If you were to sample some distance away from the questioned location, would you find similar material? That is, is the material with its unique properties common over a wide area? We need much more information and research about the occurrence and distribution of unique earth materials over short distances. However, this question is normally

answered by a study of samples collected away from the immediate scene in question and a professional choice to focus on those properties that are known to change rapidly.

2. Is there another place on earth where an exactly similar sample *might* exist? In most cases, the answer can only be yes. Although in some circumstances geologic evidence is highly individual, generally it must be discussed only in terms of similarity and dissimilarity, comparability or lack thereof. Soil samples approach individuality when several methods are used, several different soils are found that compare, or unusual minerals or combinations of minerals are found.

The experienced forensic geologist will have developed a set of procedures appropriate to the types of earth materials normally encountered.

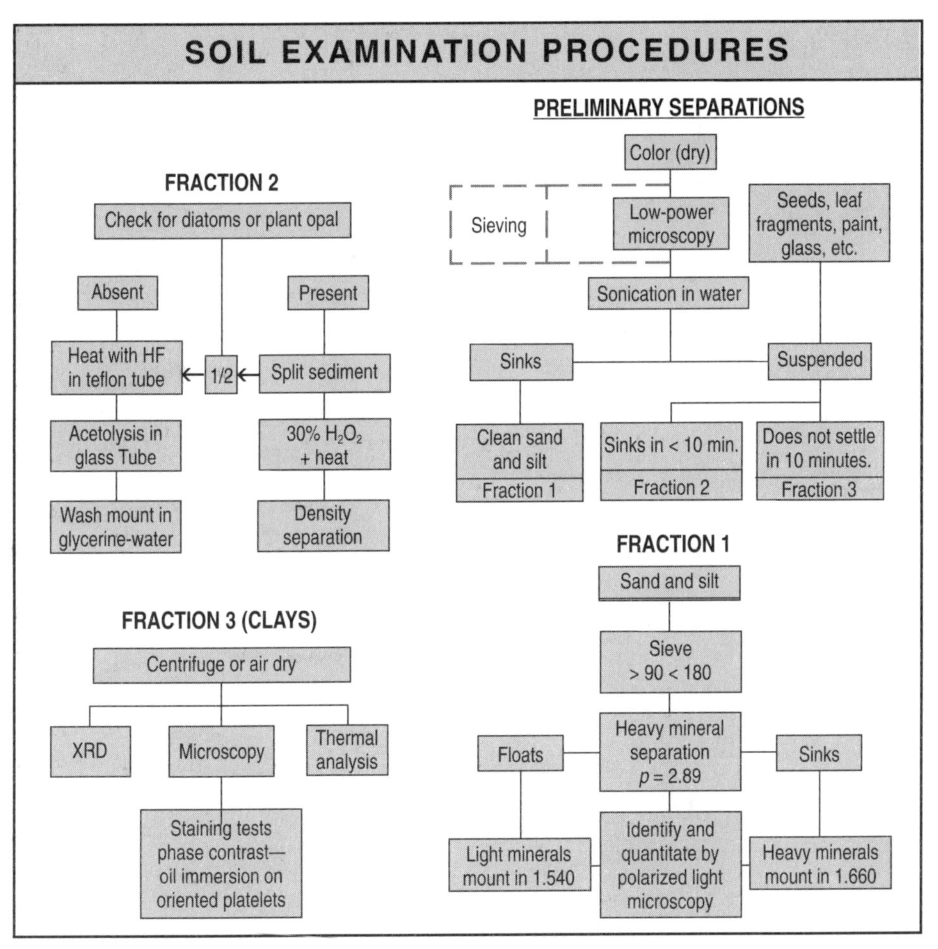

Examination sequence developed by Skip Palenik for teaching purposes

—COURTESY OF SKIP PALENIK, MICROTRACE, INC.

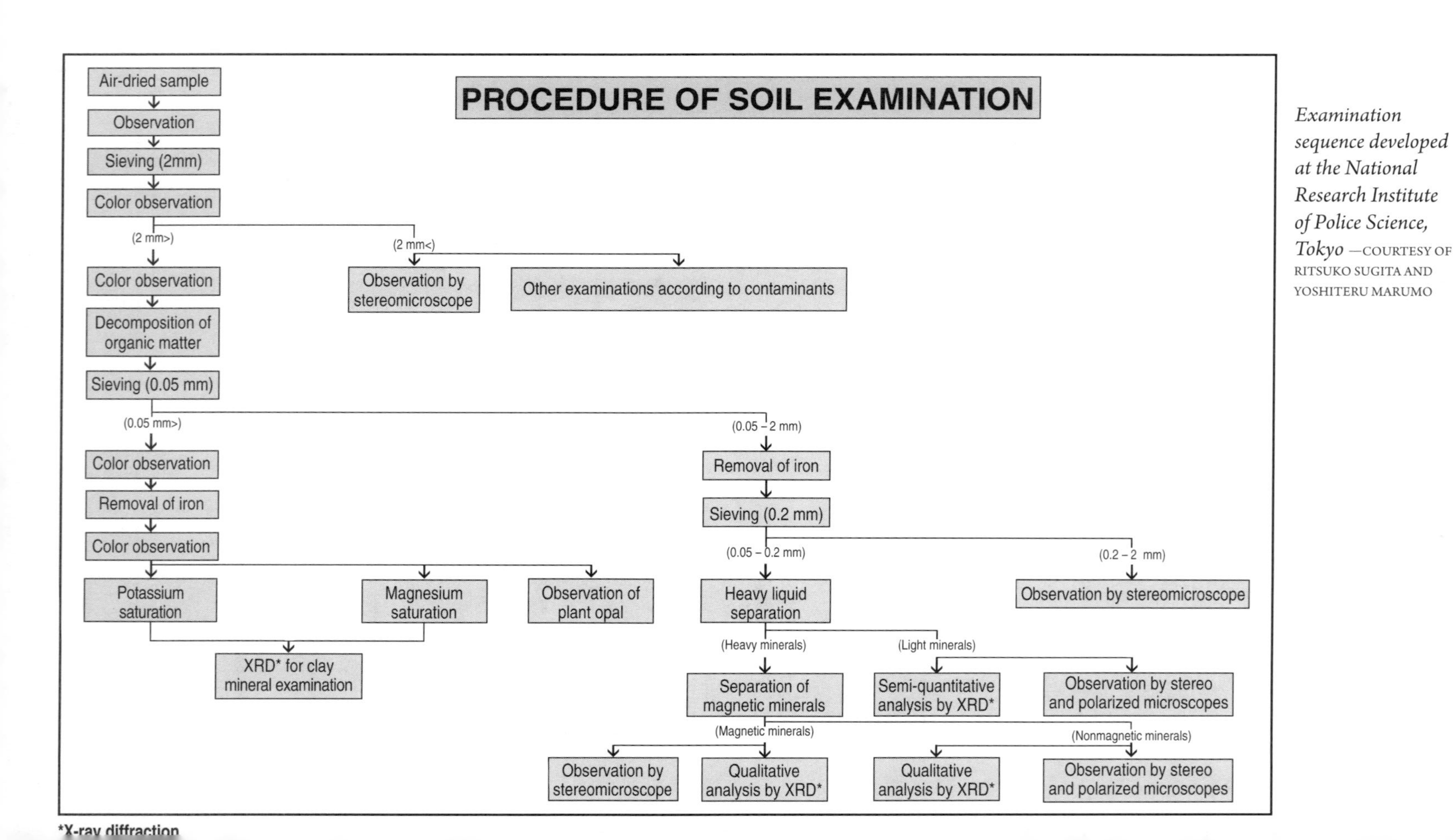

Examination sequence developed at the National Research Institute of Police Science, Tokyo —COURTESY OF RITSUKO SUGITA AND YOSHITERU MARUMO

However, the following sequence of procedures is suggested as a working model.

The sample is first examined for color, using natural light or color-determining instrumentation. Particles of various size grades may also be examined for color. Grain size distribution is determined if sufficient sample is available. The whole sample is studied under the stereo binocular microscope. Unusual particles of possible value to other specialists, such as hair, fiber, paint, and plastics are removed. The smallest size grades are saved for possible study by X-ray diffraction, scanning electron microscope, chemical analysis, or other instrumental analysis. The scientist studies the coarser material under the stereo binocular microscope, identifying and counting the kinds of particles. It may also be desirable at this stage to study thin sections or grain mounts with the petrographic microscope. Alternatively, other visual, chemical, and physical methods such as scanning electron microscopy or X-ray diffraction may be used. Each stage in the examination suggests the appropriate next step until the scientist is satisfied that the samples compare or do not compare and is prepared to defend that judgment.

Regrettably, anyone who has worked in forensic geology is familiar with testimony involving earth materials that was based on inadequate study. These examples of poor evidence have been presented for both the defense and the prosecution, and in both civil and criminal matters. There is no excuse for the submission of evidence that relies on outmoded or inadequate methods or ideas. In the application of geology to oil prospecting, we use few of the methods in general use in the 1920s. Just as the search for oil has changed, so must geoforensic work. All forensic laboratories should seek to match the work performed at the finest laboratories. The search for justice demands our best.

The Future of Forensic Geology

The last several years have witnessed a tremendous increase in both the quantity and quality of forensic examination of soil and other earth materials. The future of the art will depend on how well we address a series of issues including the following:

1. New methods are being developed to take advantage of the discriminating power inherent in earth materials. The development of X-ray diffraction techniques that provide quantitative data on the mineral composition of samples is an important direction for research.

2. Considerable effort must be devoted to defining appropriate sampling methods, communicating the potential evidential value of earth

materials to law enforcement personnel, and training those who collect samples. Unless the evidence is collected, there will be no forensic examination.

3. There is a tremendous need for studies that attempt to demonstrate the diversity and distribution of soil types. Ideally the information should be available in an electronic database and widely distributed.

4. There must be a continuing effort at all levels to improve the qualifications of examiners in forensic geology. Ongoing education could include university course work, meaningful on-the-job experience, and expanded instruction from major national forensic research organizations, including the National Academy of the Federal Bureau of Investigation. In an example of knowledge that is becoming overlooked, many geology departments today no longer teach use of the petrographic microscope for mineral identification. It is assumed students will learn the method on their own if they wish. Thus students graduating with geology degrees have neither familiarity with the instrument nor the background and experience to become skilled microscopists.

5. New instruments such as the scanning electron microscope, which provide increasingly detailed measurements and observations, make it possible to discriminate between individual grains. At that level, however, you may lose the ability to say that two samples have a common source. The whole concept of comparison is eliminated and the evidential value is lost. To avoid this problem, the forensic geologist must choose methods that provide the maximum discriminating power between samples without falsely excluding samples that are in fact associated.

The challenge for forensic geology, as with all scientific evidence, lies with education. Investigators and evidence collectors must be educated on how earth materials can make a major contribution to justice when properly collected, properly studied, and properly presented in a judicial setting. The most skillful and objective forensic scientists must produce quality evidence to serve the cause of justice. Our system of justice is still run by people, people who are human and therefore fallible, and in many cases people who are trained advocates. Those whom the courts honor by allowing the privilege of expressing an opinion—the expert scientific witnesses—must rise to the highest standards in the production of scientific evidence. If they don't, the advocates may find ways to remove the privilege, returning us to a legal world populated only by human witnesses reciting their stories from studied memory.

Glossary

These terms have been modified from *Glossary of Geology*, Second Edition, edited by Robert L. Bates and Julia A. Jackson, published by the American Geological Institute, 1980.

air-dry. The state of dryness (of a soil) at equilibrium with the moisture content in the surrounding atmosphere.

amber. A very hard, brittle, usually yellowish to brownish, translucent or transparent fossil resin derived from coniferous trees, which frequently encloses insects or other organisms.

amethyst. A transparent to translucent, purple, purple red, reddish purple, bluish violet, or pale violet variety of crystalline quartz.

amorphous. Said of a mineral or other substance lacking crystalline structure.

amphibole. A group of dark, rock-forming ferromagnesian silicate minerals closely related in crystal form and composition.

anisotropic. Having physical properties that vary when measured in different directions.

anorthite. A white, grayish, or reddish triclinic mineral of the plagioclase feldspar group.

anthracite. Coal of the highest metamorphic rank, in which the fixed-carbon content is between 92 and 98 percent.

arenaceous. Said of a sediment or sedimentary rock consisting wholly or in part of sand-sized fragments.

argillaceous. Pertaining to, largely composed of, or containing clay-sized particles or clay minerals.

arkose. A feldspar-rich, typically coarse-grained sandstone, commonly pink or reddish to pale gray or buff, composed of angular to subangular grains that may be either poorly or moderately well sorted, usually derived from the rapid disintegration of granite or granitic rocks.

asbestos. A commercial term applied to a group of highly fibrous silicate minerals that readily separate into long, thin, strong fibers of sufficient flexibility to be woven.

basalt. A dark to medium-dark colored extrusive igneous rock.

bentonite. A soft rock composed of clay minerals of the montmorillonite group and colloidal silica. It swells in water and is used in oil drilling and for sealing ponds.

biotite. A dark-colored mineral of the mica group.

black sand. An alluvial or beach sand consisting predominantly of grains of heavy, dark minerals or rocks.

bog iron ore. A general term for a soft, spongy, and porous deposit of impure hydrous iron oxides formed in bogs, marshes, swamps, peat mosses, and shallow lakes by precipitation from iron-bearing waters.

braided stream. A stream that divides into or follows an interlacing or tangled network of several small branching and reuniting shallow channels separated from each other by branch islands or channel bars, resembling in design the strands of a complex braid.

breccia. A coarse-grained clastic rock composed of large, angular, broken rock fragments that are cemented together in a fine-grained matrix.

calcite. A common rock-forming mineral, usually white, colorless, or pale shades of gray, yellow, and blue.

carbon-14 dating. A method of determining age in years by measuring the concentration of carbon-14 remaining in an organic material, usually formerly living matter.

carbonate rock. A rock consisting chiefly of carbonate minerals, such as limestone, dolomite, or carbonatite.

chalcedony. A cryptocrystalline variety of quartz.

chalk. A soft, pure, earthy, fine-textured, usually white to light gray or buff limestone of marine origin, consisting almost wholly of calcite.

chert. A hard, extremely dense or compact, dull to semivitreous, cryptocrystalline sedimentary rock, consisting dominantly of cryptocrystalline silica.

clay mineral. One of a complex and loosely defined group of finely crystalline, metacolloidal, or amorphous hydrous silicates essentially of aluminum.

coal. A readily combustible rock containing more than 50 percent by weight and more than 70 percent by volume of carbonaceous material including inherent moisture, formed from compaction and induration of variously altered plant remains similar to those in peat. Differences in the kinds of plant materials, in degree of metamorphism, and in the range of impurity are characteristics of coal and are used in classification.

colluvium. A deposit of rock fragments and soil material accumulated at the base of steep slopes as a result of gravitational action.

conglomerate. A coarse-grained, clastic sedimentary rock composed of rounded to subangular fragments larger than 2 millimeters in diameter set in a fine-grained matrix of sand, silt, or any of the common natural cementing materials.

contact. A plane or irregular surface between two different types or ages of rocks.

creep. The slow, gradual, more or less continuous, nonrecoverable deformation sustained by ice, soil, and rock materials under gravitational body stresses.

cross-bedding. An internal arrangement of the layers in a stratified rock, characterized by minor beds or laminae inclined more or less regularly in straight sloping lines or concave forms at various angles.

cryptocrystalline. Said of the texture of a rock consisting of or having crystals that are too small to be recognized and separately distinguished under the ordinary microscope.

delta. The low, nearly flat, alluvial tract of land deposited at or near the mouth of a river, commonly forming a triangular or fan-shaped plain.

diabase. In the United States, an intrusive rock whose main components are labradorite and pyroxene.

diatom. A microscopic, single-celled plant growing in marine or fresh water. Diatoms secrete siliceous skeletons of a great variety of forms that may accumulate in sediments in enormous numbers.

diatomaceous earth. A white, yellow, or light-gray siliceous earth composed predominantly of the opaline skeletons of diatoms.

dike. A tabular igneous intrusion that cuts across the planar structures of the surrounding rock.

dolomite. A common rock-forming rhombohedral mineral. It is white, colorless, or tinged yellow, brown, pink, or gray.

dowsing. The purported art or practice of locating groundwater, mineral deposits, or other objects by means of a divining rod or a pendulum.

drift. A general term applied to all rock material transported by a glacier and deposited by or from the ice or by running water emanating from a glacier.

drilling mud. A heavy suspension, usually in water but sometimes in oil, used in rotary drilling, consisting of various substances in a finely divided state.

dripstone. A general term for any cave deposit of calcium carbonate or other mineral formed by dripping water, including stalactites and stalagmites.

dust. Dry, solid organic or inorganic matter consisting of clay- and silt-sized earthy particles so finely divided that they can readily be lifted and carried considerable distances in suspension by turbulent eddies in the wind.

emery. A dark, granular, impure variety of corundum, which contains varying amounts of iron oxides and is used in the form of coarse or fine powder or grains as an abrasive for polishing and grinding.

eolian. Pertaining to the wind.

erratic. A relatively large rock fragment, often lithologically different from the bedrock on which it lies, that has been transported by glacial or floating ice, sometimes a considerable distance, from its place of origin.

extrusive. Said of igneous rock that has been ejected onto the surface of the earth, such as lava flows and volcanic ash.

fault. A surface or zone of rock fracture along which there has been displacement, from a few centimeters to a few kilometers in scale.

feldspar. A group of abundant rock-forming minerals. Feldspars are the most widespread of any mineral group and constitute 60 percent of the earth's crust.

flagstone. A hard sandstone, usually micaceous and fine-grained, that splits readily and uniformly along bedding planes or joints into large, thin, flat slabs suitable for making pavements or covering the side of a house.

flint. A mineral name for a massive, very hard, somewhat impure variety of chalcedony, usually black or gray.

fossil. Any remain, trace, or imprint of a plant or animal that has been preserved by natural processes in the earth's crust since some past geologic time.

gabbro. A group of dark-colored, basic intrusive igneous rocks.

garnet. A brittle, transparent to subtransparent mineral, most commonly dark red, with a vitreous luster and no cleavage.

gemstone. Any mineral, rock, or other natural material that, when cut and polished, has the necessary beauty and durability or hardness for use as a personal adornment or other ornament.

geophysical survey. The use of one or more geophysical techniques in geophysical exploration, such as earth currents, electrical, infrared, heat flow, magnetic, radioactivity, and seismic.

glacier. A large mass of ice formed, at least in part, on land by the compaction and recrystallization of snow, and slowly creeping downslope or outward in all directions due to its own weight.

gneiss. A foliated rock formed by regional metamorphism in which bands or lenticles of granular minerals alternate with bands and lenticles in which minerals having flaky or elongate prismatic habits predominate.

granite. A coarse-grained plutonic rock in which quartz constitutes 10 to 50 percent of the felsic components.

graphite. A hexagonal, naturally occurring crystalline form of carbon dimorphous with diamond. It is opaque, lustrous, very soft, greasy to the touch, and iron black to steel gray in color, occurring as crystals or flakes, scales, laminae, or grains.

groundwater. That part of the subsurface water that is in the zone of saturation, including underground streams.

gypsum. A widely distributed mineral consisting of hydrous calcium sulfate. It is white or colorless when pure, but commonly has tints of gray, red, yellow, blue, or brown.

halite. A native salt, occurring in massive granular, compact, or cubic crystalline forms.

hardpan. A general term for a relatively hard, impervious, and often clayey layer of soil lying at or just below the surface, produced as a result of cementation of soil particles by precipitation of relatively insoluble materials such as silica, iron oxide, calcium carbonate, and organic matter, offering exceptionally great resistance to digging or drilling, and permanently hampering root penetration and downward movement of water.

hornblende. The most common mineral of the amphibole group, commonly black, dark green, or brown, and occurring in distinct monoclinic crystals.

ice age. A loosely used synonym of *glacial epoch,* or time of extensive glacial activity; specifically, the latest of the glacial epochs known as the Pleistocene.

igneous. Said of a rock or mineral that solidified from molten or partly molten material called magma.

insoluble residue. The material remaining after a more soluble part of a rock has been dissolved in hydrochloric acid or acetic acid.

isobath. In oceanography, a line on a map or chart that connects points of equal water depth.

isopach. A line drawn on a map through points of equal thickness of a designated stratigraphic unit or group of stratigraphic units.

isotropic. Said of a medium whose properties are the same in all directions.

joint. A surface of actual or potential fracture or parting in a rock, without displacement.

karst. A type of topography that is formed over limestone, dolomite, or gypsum by dissolution, and that is characterized by closed depressions or sinkholes, caves, and underground drainage.

lacustrine. Pertaining to, produced by, or formed in a lake or lakes.

lava. A general term for molten extrusive; also, for the rock that is solidified from it.

limestone. A sedimentary rock consisting chiefly of calcium carbonate, primarily in the form of the mineral calcite, containing more than 95 percent calcite and less than 5 percent dolomite.

loess. A widespread, homogeneous, commonly nonstratified, porous, friable, unconsolidated but slightly coherent fine-grained wind deposit. It is believed to be windblown dust of Pleistocene age.

luster. The reflection of light from the surface of a mineral described by its quality and intensity.

macrofossil. A fossil large enough to be studied without the aid of a microscope.

magma. Naturally occurring molten rock material generated within the earth.

marble. A metamorphic rock consisting predominantly of fine- to coarse-grained recrystallized calcite and/or dolomite.

meander. One of a series of somewhat regular, freely developing sinuous curves, bends, loops, turns, or windings in the course of a stream.

metamorphic rock. Any rock derived from preexisting rocks by mineralogical, chemical, and/or structural changes, essentially in the solid state, in response to marked changes in temperature, pressure, shearing stress, and chemical environment at depth in the earth's crust.

mica. A group of silicate minerals with one perfect cleavage.

microcrystalline. Said of the texture of a rock consisting of or having crystals that are small enough to be visible only under a microscope.

Mohs scale. A standard of ten minerals by which the hardness of a mineral may be rated.

moraine. A mound, ridge, or other distinct accumulation of unsorted, unstratified glacial drift, predominantly till, deposited chiefly by direct action of glacier ice in a variety of topographic landforms controlled by the surface on which the drift lies.

mud. A slimy, sticky, or slippery mixture of water and finely divided particles of solid or earthy material, with a consistency varying from semifluid to that of a soft and plastic sediment.

Munsell color system. A system of color classification that is applied in geology to the colors of rocks and soils.

nodule. A small, hard, and irregular, rounded, or tuberous body of a mineral or mineral aggregate, normally having a warty or knobby surface and no internal structure, and usually exhibiting a contrasting composition from and greater hardness than the sediment or rock matrix in which it is embedded.

obsidian. A black or dark-colored volcanic glass.

oil shale. A kerogen-bearing, finely laminated brown or black shale that will yield liquid or gaseous hydrocarbons on distillation.

opal. An amorphous form of silica containing varying proportions of water and occurring in nearly all colors.

outwash. Stratified detritus removed or "washed out" from a glacier by melt-water streams and deposited in front of or beyond the terminal moraine or the margin of an active glacier.

palynology. A branch of science concerned with the study of the pollen of seed plants and spores of other plants, whether living or fossil, including their dispersal.

particle size. The effective diameter of a particle measured by sedimentation, sieving, or micrometric methods.

peat. Unconsolidated soil material consisting largely of undecomposed or slightly decomposed organic matter accumulated under conditions of excessive moisture.

pegmatite. An exceptionally coarse-grained igneous rock, with interlocking crystals, usually found as irregular dikes, lenses, or veins.

petrology. The branch of geology dealing with the origin, occurrence, structure, and history of rocks, especially igneous and metamorphic rocks.

phi grade scale. A statistical device to permit the direct application of conventional statistical practices to sedimentary data.

placer. A superficial mineral deposit formed by mechanical concentration of mineral particles from weathered debris.

plagioclase. One of the most common rock-forming minerals, which have characteristic twinning and often display zoning.

porphyry. An igneous rock of any composition that contains conspicuous crystals in a fine-grained groundmass.

pumice. A light-colored, vesicular glassy rock commonly having the composition of a rhyolite. It is often sufficiently buoyant to float on water and is economically useful as a lightweight aggregate and as an abrasive.

pyrite. A common, pale bronze or brass yellow isometric mineral.

pyroclastic rock. A rock that is composed of materials fragmented by volcanic explosion, characterized by a lack of sorting.

quartz. Crystalline silica, an important rock-forming mineral. Next to feldspar, it is the most common mineral.

quartzite. *Metamorphic:* A metamorphic rock consisting mainly of quartz and formed by recrystallization of sandstone or chert by either regional or thermal metamorphism. *Sedimentary:* A very hard but unmetamorphosed sandstone consisting chiefly of quartz grains that have been so completely and solidly cemented with secondary silica that the rock breaks across or through the individual grains rather than around them.

radiometric dating. Calculating an age in years for geologic materials by measuring the presence of a short-life radioactive element or by measuring the presence of a long-life radioactive element plus its decay product.

red beds. Sedimentary strata deposited in a continental environment, composed largely of sandstone, siltstone, and shale with locally thin units of conglomerate, limestone, or marl, and predominantly red in color due to the presence of ferric oxide usually coating individual grains.

refractometer. An apparatus for measuring the indexes of refraction of a substance, either solid or liquid.

rhinestone. An inexpensive and lustrous imitation of diamond, consisting of glass that has been backed with a thin leaf of metallic foil to simulate the brilliancy of a diamond.

rhyolite. A group of extrusive igneous rocks, generally porphyritic and exhibiting flow texture, with visible crystals of quartz and alkali feldspar in a glassy to cryptocrystalline groundmass.

riparian. Pertaining to or situated on the bank of a body of water, especially of a watercourse such as a river.

ripple mark. An undulatory surface or surface sculpture consisting of alternating, subparallel, usually small-scale ridges and hollows of primary origin.

rock salt. Coarsely crystalline halite occurring as a massive, fibrous, or granular aggregate, and constituting a nearly pure sedimentary rock that may occur in domes or plugs or as extensive beds resulting from evaporation of saline water.

sandstone. A medium-grained, clastic sedimentary rock composed of abundant rounded or angular fragments of sand set in a fine-grained matrix and more or less firmly united by a cementing material.

schist. A strongly foliated crystalline rock, formed by dynamic metamorphism, that can be readily split into thin layers or flakes due to the well-developed parallelism of more than 50 percent of the minerals present, particularly those of elongate prismatic habit.

seismic. Pertaining to an earthquake or earth vibration, including those that are artificially induced.

shale. A fine-grained, indurated, detrital sedimentary rock formed by the consolidation of clay, silt, or mud and characterized by finely stratified structure and/or fissility that is approximately parallel to the bedding and that is commonly most conspicuous on weathered surfaces.

sill. A tabular igneous intrusion that parallels the planar structure of the surrounding rock.

siltstone. An indurated silt having the texture and composition but lacking the fine lamination or fissility of shale.

slate. A compact, fine-grained metamorphic rock formed from such rocks as shale and volcanic ash, which possesses the property of fissility along planes independent of the original bedding, whereby they can be divided into plates that are lithologically indistinguishable.

soil horizon. A layer of soil or soil material approximately parallel to the land surface and differing from adjacent genetically related layers in physical, chemical, and biological properties or characteristics, such as color, structure, texture, consistency, kinds and numbers of organisms present, and degree of acidity or alkalinity.

sorting. The dynamic process by which sedimentary particles having some particular characteristic are naturally selected and separated from associated but dissimilar particles by the agents of transportation.

spinel. A mineral having a great hardness and usually forming octahedral crystals. It varies widely in color depending on the chemical composition, and sometimes is used as a gemstone.

spit. A small point or low tongue or narrow embankment of land commonly consisting of sand or gravel deposited by long-shore drifting and having one end attached to the mainland and the other terminating in open water.

stalactite. A conical or cylindrical deposit that hangs from the roof of a cave. It is deposited by dripping water and is usually composed of calcium carbonate, but may also be formed of metallic carbonates.

stalagmite. A conical deposit that is developed upward from the floor of a cave by the action of dripping water. It is usually composed of calcium carbonate, but may be composed of metallic carbonates.

staurolite. A brownish to black orthorhombic mineral. It is often twinned so as to resemble a cross.

stratigraphy. The branch of geology that deals both with the definition and description of major and minor natural divisions of rocks available for study in outcrop or from subsurface, and with the interpretation of their significance in geologic history.

streak. The color of a mineral in its powdered form, usually obtained by rubbing the mineral on a streak plate and observing the mark it leaves.

talc. An extremely soft, whitish, greenish, or grayish monoclinic mineral. It has a characteristic soapy or greasy feel and can be cut with a knife.

talus. Rock fragments of any size or shape derived from and lying at the base of a cliff or a very steep, rocky slope.

till. Unsorted and unstratified drift, generally unconsolidated and deposited directly by and underneath a glacier, without subsequent reworking by water from the glacier.

trace element. An element that is not essential in a mineral but that is found in small quantities in its structure or adsorbed on its surfaces.

travertine. A hard, dense, finely crystalline compact or massive but often concretionary limestone of white, tan, or cream color, often having a fibrous or concentric structure and splintery fracture.

varve. A sedimentary bed or lamina or sequence of laminae deposited in a body of still water within one year's time.

vesicle. A cavity of variable shape in a lava, formed by the entrapment of a gas bubble during solidification of the lava.

well log. A record of the measured or computed characteristics of the rock section in a well, showing such information as resistivity, radioactivity, spontaneous potential, and acoustic velocity as a function of depth.

zone of saturation. A subsurface zone in which all the interstices are filled with water under pressure greater than that of the atmosphere.

Additional Resources

Bisbing, R. E. 1989. Clues in the dust. *American Laboratory* (November):19-23.

Block, E. B. 1958. *The Wizard of Berkeley.* New York: Coward-McCann.

_______. 1979. *Science vs. Crime.* San Francisco: Cragmont Publications.

Brown, A. G. 2006. The use of forensic botany and geology in war crimes investigation in northeast Bosnia. *Forensic Science International.* 163:204–10

Chaperlin, K., and P. S. Howarth. 1983. Soil comparison by the density gradient method—A review and evaluation. *Forensic Science International* 23:161–77.

Daniels, F. 2003. *Dead Center.* Far Hills, N.J.: New Horizon Press.

Davenport, G. C. 2001. *Where Is It? Searching for Buried Bodies and Hidden Evidence.* Lakewood, Colo.: GeoForensics International.

Davenport, G. C., and others. 1990. Geoscientists and law enforcement professionals work together in Colorado. *Geotimes* (July):13–15.

Doyle, A. C. 1956. *The Complete Sherlock Holmes,* vol. 1. New York: Doubleday.

Dudley, R. J. 1975. The use of color in the discrimination between soils. *Journal of Forensic Sciences* 15:209–18.

Dudley, R. J., and K. W. Smalldon. 1978. The evaluation of methods for soil analysis under simulated scenes of crime conditions. *Forensic Science International* 12:49–60.

Edwards, H. G. M. 1999. Art works studied using IR and raman spectroscopy. In *Encyclopedia of Spectroscopy and Spectrometry,* ed. J. C. Lindon, G. E. Tranter, and J. L. Holmes. London: Academic Press.

Federal Bureau of Investigation. 2008. *FBI Handbook of Crime Scene Forensics.* New York, N.Y.: Skyhorse Publishing. Inc.

Fitzpatrick, R. W., M. D. Raven, M. Heath, and G. Rinder. 2007. How soil evidence helped solve a double murder case: A display. In the Abstracts for the Second International Conference on Criminal and Environmental Soil Forensics. Edinburgh, England.

Frenkel, 0. J. 1965. A program of research into the value of evidence from southern Ontario soils. *Proceedings of the Canadian Society of Forensic Science* 4:23.

_______. 1968. Three studies on the forensic comparison of soil samples. Paper read at 1968 meeting of the American Academy of Forensic Sciences, Chicago, Ill.

Graves, W. J. 1979. A mineralogical soil classification technique for the forensic scientist. *Journal of Forensic Sciences* 24:323–39.

Gross, H. 1893. *Handbuch für Untersuchungsrichter.* Munich.

Harrison, M., and L. J. Donnelly. 2009. Locating concealed homicide victims; developing the role of Geoforensics. In: Ritz, K., Dawson, L. & Miller, D. (eds). *Criminal and Environmental Soil Forensics.* Springer, 197–219.

Horrocks, M., and K. J. Walsch. 2001. Pollen on grass clippings: Putting the suspect at the scene of the crime. *Journal of Forensic Sciences* 46 (4):947–49.

Houck, M. 2004. *Trace Evidence Analysis.* Elsevier Academic Press.

Kirk, P. L. 1953. *Crime Investigation.* New York: Interscience Publishers.

_______. 1962. *Criminal Investigation.* Trans. J. Adam and J. Collier Adam, revised by R. L. Jackson. London: Sweet and Maxwell.

Lee, B. D., T. N. Williamson, and R. C. Graham. 2002. Identification of stolen rare palm trees by soil morphological and mineralogical properties. *Journal of Forensic Sciences* 47 (1):190–94.

Lee, H. 2002. *Cracking Cases.* Amherst, N.Y.: Prometheus Books.

Lee, H., T. Palmbach, and M. Miller. 2001. *Henry Lee's Crime Scene Handbook.* Burlington, Mass.: Academic Press.

Lindemann, J. W. 2000. Forensic geology. *The Professional Geologist* 37 (9):4–7.

Lombardi, G. 1999. The contribution of forensic geology and other trace evidence analysis to the investigation of the killing of Italian Prime Minister Aldo Moro. *Journal of Forensic Sciences* 44 (3):634–42.

_______. 2009. The Death of Countess Agusta in Portofino (Northern Italy) and the Soil from Two Mismatched Slippers. *Journal of Forensic Science* 54 (2):395–99

Marumo, Y., R. Sugita, and S. Seta. 1995. Soil as evidence in criminal investigation. Eleventh INTERPOL Forensic Science Symposium. Lyon, France.

McCrone, W. C., and J. G. Delly. 1973. *The Particle Atlas,* vols. 1–6. Ann Arbor, Mich.: Ann Arbor Science Publishers.

McCrone, W. C., D. Chartier, and R. Weiss. eds. 1998. *Scientific Detection of Fakery in Art.* Bellingham, Wash.: The International Society for Optical Engineering.

McCrone, W. C., L. Graham, and J. A. Polizzi. 1996. Christ Among the Doctors: A new Leonardo painting? *Microscope* 44 (3):119–36.

McPhee, J. 1997. *Irons in the Fire.* New York: Farrar, Straus and Giroux.

Miller, P. S. 1996. Disturbances in the soil: Finding buried bodies and other evidence using ground penetrating radar. *Journal of Forensic Sciences* 41 (4):648–52.

Moenssens, A. A., R. E. Moses, and F. E. Inbau. 1995. *Scientific Evidence in Civil and Criminal Cases,* 4th ed. New York: Foundation Press.

Murray, R. C. 1975. The geologist as private eye. *Natural History Magazine,* February, 22–26.

_______. 1976. Soil and rocks as physical evidence. *Law and Order,* July, 36–40.

_______. 1988. Forensic geology—100 years. *Microscope* 36 (4):303–8.

Murray, R. C., and J. C. F. Tedrow. 1975. *Forensic Geology: Earth Sciences and Criminal Investigation.* New Brunswick, N.J.: Rutgers University Press.

_______. 1991. *Forensic Geology.* Englewood Cliffs, N.J.: Prentice Hall.

_______. 2010. Forensic Geology—Getting the Dirt on Crime. *International Game Warden* Summer: 32–33.

Murray, R. C., and R. Murray. 1980. Soil evidence. *Law and Order,* July, 26-28.

Ojena, S. M., and P. R. Deforest. 1972. Precise refractive index determination of the immersion method, using phase contrast microscopy and the Mettler hot stage. *Journal of Forensic Sciences* 12:315-29.

Palenik, S. J. 1979. The determination of geographical origin of dust samples. In *The Particle Atlas,* vol. 5, 2nd ed., ed. W. C. McCrone and others. Ann Arbor, Mich.: Ann Arbor Science Publishers.

_______. 1993. The analysis of dust traces. *Proceedings of the International Symposium on the Forensic Aspects of Trace Evidence.* Washington, D.C.: Government Printing Office.

Petraco, N., and T. Kubic. 2000. A density gradient technique for use in forensic soil analysis. *Journal of Forensic Sciences* 45 (4):872–73.

Pye, K. 2007. *Geological and Soil Evidence.* CRC Press.

Pye, K., and D. J. Croft. 2004. *Forensic Geoscience.* Geological Society of London Special Publication 232.

Rapp, J. S. 1987. Forensic geology and a Colusa County murder. *California Geology,* 147-53.

Ritz, K., and others. 2009. *Criminal and Environmental Soil Forensics.* Springer

Ruffell, A., and J. McKinley. 2008. *Geoforensics.* John Wiley and Sons.

Saferstein, R. 2011. *Criminalistics: An Introduction to Forensic Science,* 10th ed. Upper Saddle River, N.J.: Prentice Hall.

Saferstein, R., ed. 2002. *Forensic Science Handbook,* 2nd ed. Upper Saddle River, N.J.: Prentice Hall. See especially the sections on glass (R. D. Koons, J. Buscaglia, M. Bottrell, and T. Miller) and soils (R. C. Murray and L. P. Solebello).

Smale, D. 1973. The examination of paint flakes, glass and soils for forensic purposes, with special reference to electron probe microanalysis. *Journal of Forensic Science Society* 13:5.

Smale, D., and N. A. Trueman. 1969. Heavy mineral studies as evidence in a murder case in outback Australia. *Journal of Forensic Science Society* 9:3-4.

Stam, M. 2002. The dirt on you. *California Association of Criminalist's Newsletter* 2:8–11.

Stanley, E. A. 1992. Application of palynology to establish the provenance and travel history of illicit drugs. *Microscope* 40:149–52.

Strongman, K. B. 1992. Forensic applications of ground penetrating radar. In *Ground Penetrating Radar,* ed. J. Pilon, 203–211. Geological Survey of Canada, paper 90-4.

Sugita, R., and Y. Marumo. 1996. Validity of color examination for forensic soil identification. *Forensic Science International* 83:201–10.

Thornton, J. I. 1975. The use of an agglomerative numerical technique in physical evidence comparison. *Journal of Forensic Sciences* 20:693–700.

_______. 1986. Forensic soil characterization. *Forensic Science Progress 1.* Heidelberg: Springer-Verlag.

Thornton, J. I., and F. Fitzpatrick. 1978. Forensic science characterization of sand. *Journal of Forensic Sciences* 20:460–75.

Thornton, J. I., and A. D. McLaren. 1975. Enzymatic characterization of soil evidence. *Journal of Forensic Science* 20:674–92.

Thorwald, J. 1967. *Crime and Science: The New Frontier in Criminology.* New York: Harcourt Brace Jovanovich.

Wanogho, S., G. Gettinby, B. Caddy, and J. Robertson. 1989. Determination of particle size distribution of soils in forensic science using classical and modern instrumental methods. *Journal of Forensic Sciences* 34 (4):823–35.

Wehrenberg, J. P. 1988. *Manual for Forensic Mineralogy, Short Course.* Missoula, Mont.: Northwest Association of Forensic Scientists.

The Roadside Geology series of books for individual states is an excellent source for local geology. The series is published by Mountain Press, Missoula, Montana.

The Federal Bureau of Investigation publishes *Forensic Science Communications* online at www.fbi.gov/about-us/lab/forensic-science-communications. This publication provides new and interesting articles on forensic science. For example, see: Max M. Houck, Statistics and trace evidence: The tyranny of numbers, vol. 1, no. 3 (October 1999).

The International Union of Geological Sciences (IUGS) Initiative on Forensic Geology was established at UNESCO headquarters in Paris, France, on February 22, 2011. The initiative aims "to develop forensic geology internationally and promote its applications." Their Web site is www.forensicgeology international.org/

For up-to-date information on the U.S. Supreme court decision in Daubert and that decision's impact on expert testimony see the article "Preparing for a Daubert Hearing" by Andre A. Moenssens at www.forensic-evidence.com/site/ID/ID_FBI.html

For more information on the Busang gold fraud see the article by Andrew Alden at: geology.about.com/cs/mineralogy/a/aa042097.htm

A major source of information and maps is the Web site of the U.S. Geological Survey at www.usgs.gov. See especially their Earth Resource Observation Systems (EROS) Data Center at http://edc.usgs.gov.

Index

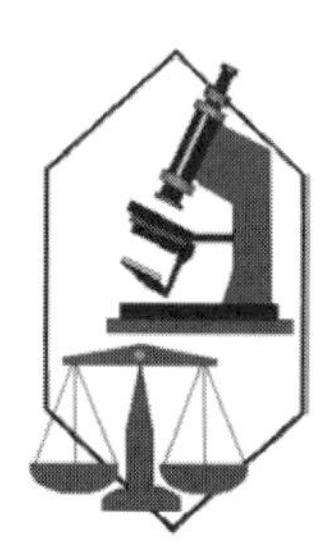

Raymond C. Murray received a Ph.D. in geology from the University of Wisconsin and later chaired the geology department at Rutgers University. He served as vice president for research and professor of geology at the University of Montana for twenty years, until his retirement in 1996. Murray is best known for his pioneering book *Forensic Geology*, coauthored with John C. F. Tedrow. Murray has worked to help establish proper methods for examination and analysis of soils and related types of evidence, and has testified as an expert witness. He has visited, worked with, and lectured at government and private crime laboratories around the world. In 2010, in recognition of his pioneering efforts in the field, Murray was presented with the prestigious Forensic Geoscience Group Award. More information on Murray and his work can be found at www.forensicgeology.net. He lives with his wife, Maureen Fleming, in Missoula, Montana.